Lecture Notes in Mathematics

Edited by A. Dold and B. Eckmann

999

Chris Preston

Iterates of Maps on an Interval

Springer-Verlag
Berlin Heidelberg New York Tokyo 1983

Author

Chris Preston
Universität Bielefeld, USP-Mathematisierung
4800 Bielefeld, Federal Republic of Germany

AMS Subject Classifications (1980): 26 A 18, 54 H 20

ISBN 3-540-12322-9 Springer-Verlag Berlin Heidelberg New York Tokyo
ISBN 0-387-12322-9 Springer-Verlag New York Heidelberg Berlin Tokyo

Printing and binding: Beltz Offsetdruck, Hemsbach/Bergstr.
2146/3140-543210

for Os and Tam

The elegant body of mathematical theory pertaining to linear systems (Fourier analysis, orthogonal functions, and so on), and its successful application to many fundamentally linear problems in the physical sciences, tends to dominate even moderately advanced University courses in mathematics and theoretical physics. The mathematical intuition so developed ill equips the student to confront the bizarre behaviour exhibited by the simplest of discrete nonlinear systems, such as the equation $x_{n+1} = ax_n(1-x_n)$ *. Yet such nonlinear systems are surely the rule, not the exception, outside the physical sciences.*

Bob May in *Simple mathematical models with very complicated dynamics* in *Nature*, Vol. 261, June 1976.

These are some notes on the iterates of maps on an interval, which we hope can be understood by anyone who has had a basic course in (one-dimensional) real analysis. The main reason for writing this account is as an attempt to make the very beautiful mathematics behind the *bizarre behaviour exhibited by the simplest of discrete nonlinear systems* accessible to as wide an audience as possible.

Parts of these notes have appeared as Volumes 34 and 37 in the series: *Materialien des Universitätsschwerpunktes Mathematisierung* from the Universität Bielefeld, and I would like to thank the USP Mathematisierung for their support. Thanks also to David Griffeath for some pertinent comments on the text and to Bob May and Alister Mees for getting me interested in this subject.

Bielefeld   Chris Preston

October 1982

ITERATES OF MAPS ON AN INTERVAL - CONTENTS

1. INTRODUCTION

Suppose we are studying some physical or biological system on which we make measurements at regular intervals (say once a year or every ten seconds). If we are just measuring a single quantity then the n th. measurement can be represented by a real number x_n. The data we thus obtain is a sequence $x_0, x_1, \ldots, x_m$ of real numbers, where of course $m+1$ is the number of measurements which are made. A very simple mathematical model of such a system is obtained by assuming that x_{n+1} is only a function of x_n, and that this function does not depend on n. That is, we assume there is a function $f : X \to X$ so that $x_{n+1} = f(x_n)$ for all $n \geq 0$, where $X \subset \mathbb{R}$ represents the set of possible values which can be registered by our measuring apparatus. Thus if a measurement at time 0 gave a value $x \in X$ then the model predicts that a measurement at time n would register a value of $f^n(x)$, where $f^n : X \to X$ is given inductively by $f^0(x) = x$, $f^1(x) = f(x)$ and $f^n(x) = f(f^{n-1}(x))$. With such a model we are therefore interested in the iterates $\{f^n\}_{n\geq 0}$ of the function f.

Typical functions which have been used (for example as models in population biology; see, for instance, May (1976)) are the functions $f_\mu : [0,1] \to [0,1]$ (with $0 < \mu \leq 4$) and $g_r : [0,1] \to [0,1]$ (with $r > 1$) given by $f_\mu(x) = \mu x(1-x)$ and $g_r(x) = rx \exp(1-rx)$. The important feature common to the functions in these two families is that they all look something like the picture at the top of the next page. More precisely, each of the functions has a unique maximum in $(0,1)$ and is strictly increasing (resp. strictly decreasing) to the left (resp. right) of this point. Thus the f_μ and g_r are elements of the set S

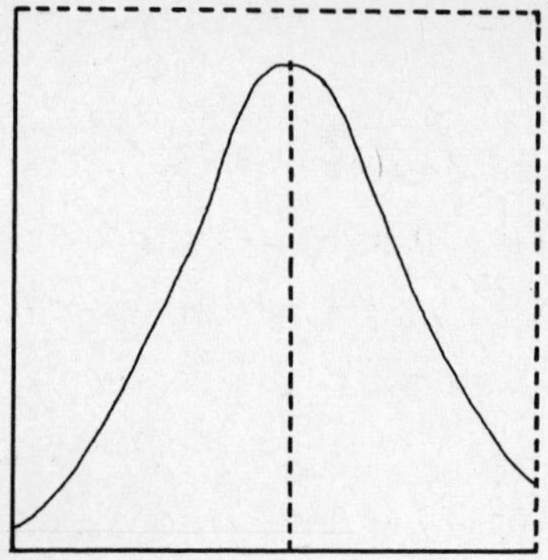

consisting of those continuous functions $f : [0,1] \to [0,1]$ for which there exists $\varphi \in (0,1)$ such that f is strictly increasing on $[0,\varphi]$ and is strictly decreasing on $[\varphi,1]$.

These notes will be concerned with the question: What kinds of behaviour can be exhibited by the iterates of a function in S ? With the help of a programmable calculator the reader can soon convince himself that for some functions in S the behaviour of the iterates is very simple, while for others it is extremely complicated. Consider, for example, the family of functions $\{f_\mu\}_{0<\mu\leqq 4}$ in S given by $f_\mu(x) = \mu x(1-x)$. For each of the five parameter values $\mu = 2.5$, 3.1 , 3.569946 , 3.8291 and 4 compute the first 10000 or so terms of the orbit $\{f_\mu^n(x)\}_{n\geqq 0}$ corresponding to a "randomly" chosen starting point $x \in [0,1]$ (where we have written and will continue to write f_μ^n instead of $(f_\mu)^n$). Repeat this several times with other "random" starting points. The following "facts" can then be observed:

<u>$\mu = 2.5$</u> . 0.6 is a fixed point of f_μ (i.e. $f_\mu(0.6) = 0.6$) and all

the points in $(0,1)$ are attracted to this point, i.e. $\lim_{n\to\infty} f_\mu^n(x) = 0.6$ for all $x \in (0,1)$. (Note however that 0 and 1 are not attracted to 0.6 since $f_\mu(0) = f_\mu(1) = 0$.)

<u>$\mu = 3.1$</u>. 0.55801.. is a periodic point with period 2 and practically all the points in $[0,1]$ are attracted to the period orbit $\{0.55801.., 0.76456..\}$. (If $f : [0,1] \to [0,1]$ then $x \in [0,1]$ is called periodic if $f^m(x) = x$ for some $m \geqq 1$; the smallest $m \geqq 1$ with this property is called the period of x. If x is periodic with period m then we say $y \in [0,1]$ is attracted to the periodic orbit $\{x, f(x), \ldots, f^{m-1}(x)\}$ if $\lim_{n\to\infty} f^{mn}(y) = f^k(x)$ for some $0 \leqq k < m$.)

In the present case "practically all" means except for the three points 0, 1 and 0.6774... ($0.6774.. = \frac{2.1}{3.1}$ is the unique fixed point of f_μ in $(0,1)$.)

<u>$\mu = 3.569946$</u>. The orbit of a randomly chosen $x \in [0,1]$ appears to be attracted to an orbit which is almost, but not quite, periodic. More precisely, the following can be observed: For $m \geqq 1$ let $x_n^m \in [0,1]$ be obtained by rounding off $f_\mu^n(x)$ to m decimal places. Then for each $m \geqq 1$ there is a periodic sequence $\{y_n^m\}_{n\geqq 0}$ (i.e. $y_{n+k}^m = y_n^m$ for some $k \geqq 1$ and for all $n \geqq 0$) such that for each randomly chosen x there exists $j \geqq 0$ with $x_{n+j}^m = y_n^m$ for all $n \geqq 0$. However, the period of the sequence $\{y_n^m\}_{n\geqq 0}$ grows very rapidly with m. (For $m = 1$ the period is 16, for $m = 2$ it is 128.) Another feature which can be seen with this value of μ is that the orbits of points near a randomly chosen x look very similar to the orbit of x. (For each

$\varepsilon > 0$ it is possible to find $\delta > 0$ so that if $|x-y| < \delta$ then $|f_\mu^n(x)-f_\mu^n(y)| < \varepsilon$ for all $n \geq 0$.)

$\underline{\mu = 3.8291}$. 0.15747.. is a periodic point with period 3 and almost all points in [0,1] are attracted to the periodic orbit {0.15747.., 0.50802.., 0.95702..} . We will later see that "almost all" can here be taken in the sense of Lebesgue measure on [0,1] , but that the situation is more complicated than in the first two cases in that there are uncountably many points which do not get attracted to this periodic orbit.

$\underline{\mu = 4}$. The orbit of a randomly chosen $x \in [0,1]$ appears to be completely "chaotic". Moreover, the orbit of x looks nothing like the orbit of a point chosen "randomly" in any small neighbourhood of x .

The above numerical examples suggest that there are at least three different kinds of behaviour which can be exhibited by the iterates of a function in $\mathcal{S}$. These are:

(1) There is a periodic orbit which attracts "almost all" of the points in [0,1] . (This case occurs for $\mu = 2.5$, 3.1 and 3.8291 .)

(2) A typical orbit appears to be completely random; there is a sensitive dependence to initial conditions. ($\mu = 4$)

(3) There is a "strange attractor" which attracts "almost all" of the points in [0,1] ; there is no sensitive dependence to initial conditions. ($\mu = 3.569946$)

These three types of behaviour will provide the key to understanding the iterates $\{f^n\}_{n\geq 0}$ of a function $f \in \mathcal{S}$. It turns out that for functions during development (developmental pathways somewhat rigid)

↳ increase in size of brain is limited

Purpose of population genetics:

「determine how evolution occurs within the framework of biological constraints」

MODELS

Model: intentional simplification of a complex situation in order to eliminate extraneous detail in order to focus attention on essentials of the situation.

exponential growth.

assumption: pop. increases by a constant factor each generation

$$N_t = (1+r) N_{t-1}, \quad r > 0.$$

$$\Rightarrow N_t = (1+r)^t N_0.$$

↳ this is ok in short term.

$$N_t - N_{t-1} = r N_{t-1}$$

$$\frac{dN_t}{dt} = r N_t, \quad N_t = N_0^{\,rt}$$

2, 5, 8, 12

$Z_{t-1} = 0 \Rightarrow Z_t = 0.$

$$Z_t = (1+r) Z_{t-1} - r Z_{t-1}^2$$

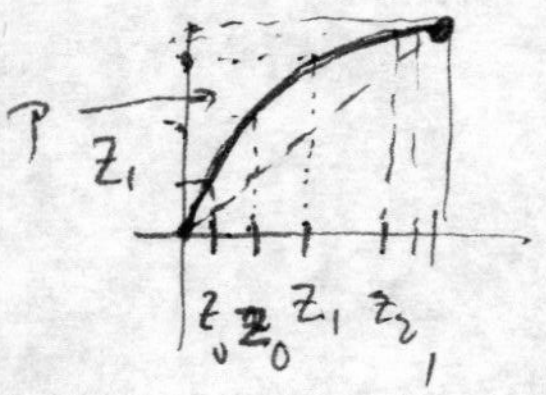

$p(x) = (1+r)x - r x^2$

$p'(x) = (1+r) - 2rx$

$= 0$

if $x = \frac{1+r}{2r}$

$1 > r > 0$

$$Z_{t+1} = r Z_t (1 - Z_t)$$

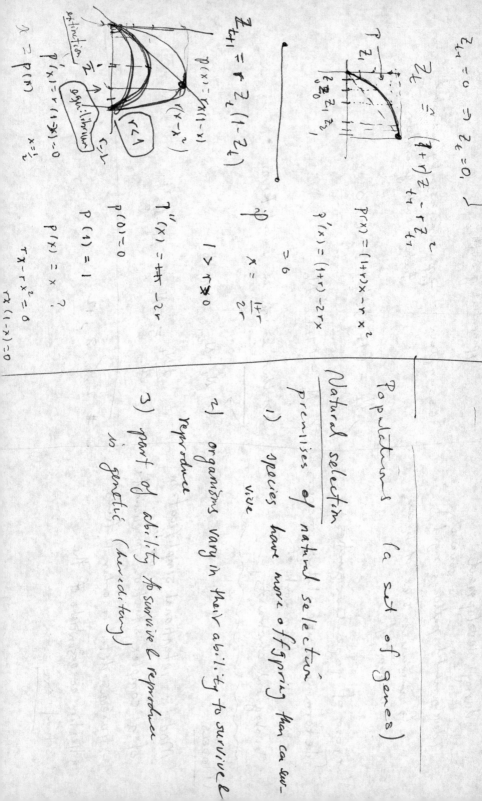

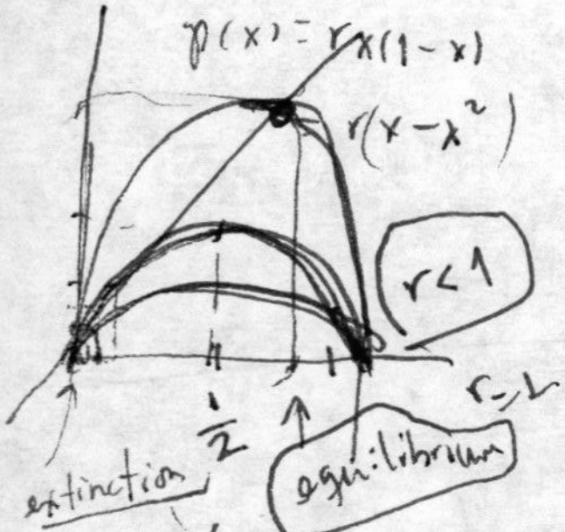

$p'(x) = r(1-2x) = 0$

$x = \frac{1}{2}$

$x = p(x)$

$p''(x) = -2r$

$p(0) = 0$

$p(1) = 1$

$p(x) = x$?

$rx - rx^2 = 0$

$rx(1-x) = 0$

Populations (a set of genes)

Natural selection

premises of natural selection

1) species have more offspring than can survive

2) organisms vary in their ability to survive & reproduce

3) part of ability to survive & reproduce is genetic (hereditary)

logistic model

$$\frac{dN}{dt} = rN(K-N)/K$$

K = carrying capacity

$$N_t - N_{t-1} = rN_{t-1}(K - N_{t-1})/K$$

$$\frac{N_t}{K} = \frac{N_{t-1}}{K} + r\frac{N_{t-1}}{K}(K - N_{t-1})/K$$

$$Z_t = \frac{N_t}{K} \qquad dZ_t = rZ_t(1-Z_t)$$

$$Z_0 = z_0$$

$$Z_t = Z_{t-1}(1 + r(1 - Z_{t-1})).$$

$$= Z_{t-1} + rZ_{t-1}(1 - Z_{t-1})$$

Lotka - Volterra Competition

competitive interactions

$$\frac{dN_t^1}{dt} = r_1 N_t^1\left(1 - \frac{N_t^1 + \alpha_{21} N_t^2}{K_1}\right)$$

$$\frac{dN_t^2}{dt} = r_2 N_t^2\left(1 - \frac{N_t^2 + \alpha_{12} N_t^1}{K_2}\right)$$

Scope of Population Genetics

evolution occurs as a result of progressive change in kinds & frequencies of genes that occur in populations,

Limits to evolution

1. alterations to codon-amino acid correspondences can not suddenly alter (resulting proteins produced by cell would be detrimental)

2. limits set by Mendelian segregation and recombination.

3 limits set by patterns of

problems in population genetics

pop. size

mating patterns

geographical dist. of individuals

mutation

migration

natural selection

via simplified experiments,

Mathematical Model. set of hypotheses that specifies mathematical relationships between measured or measurable quantities (

model must be simple enough to handle

in an important subset of $\mathcal{S}$ (which includes the families $\{f_\mu\}_{0<\mu\leq 4}$ and $\{g_r\}_{r>1}$) the three cases (1), (2) and (3) completely classify the behaviour which can occur. For a general function in $\mathcal{S}$ the situation can be more complicated, but (1), (2) and (3) are still the basic prototypes for the possible kinds of behaviour.

The mathematical results which are behind the statements in the above paragraph are due to Guckenheimer and Misiurewicz (Guckenheimer (1979), Misiurewicz (1980)). It is the object of these notes to give an elementary account of these and other results on the iterates of functions in $\mathcal{S}$. We have tried to make this account understandable to anyone who knows the basic facts about continuous and differentiable functions of one real variable (as can be found, for example, in the first five chapters of Rudin: Principles of Mathematical Analysis (1964)). In fact one of the main reasons for writing these notes is as an attempt to make some very beautiful mathematics accessible to as wide an audience as possible.

Before giving an outline of what is contained in the various sections of these notes we will make a couple of general remarks.

1. A lot of the interest in the behaviour of maps on an interval was kindled by the review article of May in *Nature* (May (1976)). The reader is strongly recommended to look at this article.

2. A second strong recommendation is to study the book by Collet and Eckmann (1980) called *Iterated maps on the interval as dynamical systems*. As its title suggests, it is concerned with much the same material as we will consider and in particular it gives an account of the results of Guckenheimer and Misiurewicz. Our justification for writing the present set of notes is that many of our proofs are simpler than the corresponding

ones in Collet and Eckmann. In any case, a second account will have served some useful purpose if it increases the number of people who are interested in the iterates of maps on an interval.

3. We only consider the behaviour of the iterates of a single function. However, in practice it is often more important to study how this behaviour changes when we vary some parameter. For example, how does the behaviour of the iterates of f_μ change as μ increases in the interval $(0,4]$? There has been a lot of interest in such questions in the last couple of years; this interest started with the discovery by Feigenbaum (Feigenbaum (1978), (1979)) that the successive bifurcations in any reasonable one-parameter family of functions from $\mathcal{S}$ exhibit a remarkable quantitative universality (in that the rates at which the bifurcations occur involve constants which are common to all such families). Unfortunately, the mathematics needed to handle these problems rigorously is way beyond what we intend to use here, and so we will not be able to study this topic. The reader is recommended to look at Hofstadter's column (*Metamagical Themas*) in the November 1981 *Scientific American*. Anyone who wants to see what kind of mathematics is involved in this area can also look at Collet, Eckmann and Lanford (1980).

4. We have made several statements involving "almost all" of the points in $[0,1]$. For most of these notes this will have a topological, rather than a measure-theoretical, meaning. In Sections 3 and 4 we will consider a set to contain "almost all" of the points in $[0,1]$ if it contains a dense open subset of $[0,1]$, and in the following sections if it contains a countable intersection of dense open subsets (i.e. if it is a residual subset of $[0,1]$ in the terminology of the Baire category theorem). The main reason for taking this approach is that it greatly simplifies a lot of the proofs; it also allows a lot of the notes to be read by someone

who has had no measure theory. In Section 9 we consider a measure-theoretic version of the main result obtained in the previous sections, and then "almost all" will mean in the sense of Lebesgue measure.

5. The fact that the functions in $\mathcal{S}$ are defined on the interval $[0,1]$ is not important. Suppose $f : [a,b] \to [a,b]$ is a continuous function for which $\xi \in (a,b)$ exists so that f is strictly increasing on $[a,\xi]$ and is strictly decreasing on $[\xi,b]$. Then we can define $g \in \mathcal{S}$ by $g(x) = (b-a)^{-1}\{f((1-x)a+xb)-a\}$ (i.e. by making a linear change of variables), and any properties we are interested in will be invariant under this transformation.

6. Our description of what happens to the iterates $\{f_\mu^n\}_{n\geqq 0}$ in the case when $\mu = 3.569946$ was not completely honest. This value of μ is only an approximation to the value of the parameter which really gives the behaviour we described. (The "correct" value of μ lies between 3.5699456 and 3.5699457.) In fact, when $\mu = 3.569946$ then there is a periodic orbit with period 23×2^{10} which attracts "almost all" of the points in $[0,1]$. (In Section 10 we will explain how the parameter value $\mu = 3.569946$ was chosen.)

7. There are many topics involving the iterates of a single function from $\mathcal{S}$ which we do not consider in these notes. Perhaps the most important concerns the existence of absolutely continuous invariant probability measures. Let $f \in \mathcal{S}$ and μ be a probability measure on $([0,1],\mathcal{B})$, where $\mathcal{B}$ denotes the Borel subsets of $[0,1]$; μ is called invariant under f if $\mu(f^{-1}(F)) = \mu(F)$ for all $F \in \mathcal{B}$. A question which has received a lot of attention recently is: Which functions in $\mathcal{S}$ have an absolutely continuous (with respect to Lebesgue measure) invariant probability measure? (μ is absolutely continuous with respect

to Lebesgue measure if $\mu(F) = 0$ whenever the Lebesgue measure of F is zero.) The main result in this direction is due to Misiurewicz (Misiurewicz (1980)); an account of this result can be found in Collet and Eckmann (1980). Another topic which we do not consider is the problem of determining whether or not two given functions from $\mathcal{S}$ are topologically conjugate. Continuous functions $f, g : [0,1] \longrightarrow [0,1]$ are said to be topologically conjugate if there exists a homeomorphism $h : [0,1] \longrightarrow [0,1]$ such that $f = h^{-1} \circ g \circ h$. If $f = h^{-1} \circ g \circ h$ then we also have $f^n = h^{-1} \circ g^n \circ h$ for each $n \geq 1$; thus if $f, g \in \mathcal{S}$ are topologically conjugate then the iterates of f and g will exhibit the same kind of topological behaviour. A well-known example of this is provided by the functions $f(x) = 4x(1-x)$ and the piecewise linear function $g(x) = \begin{cases} 2x & \text{if } 0 \leq x \leq \frac{1}{2} , \\ 2-2x & \text{if } \frac{1}{2} \leq x \leq 1 . \end{cases}$ (For these functions we have $f = h^{-1} \circ g \circ h$ with $h(x) = \frac{1}{2} + \frac{1}{\pi} \sin^{-1}(2\sqrt{x} - 1)$.) For an account of this subject the reader is again referred to Collet and Eckmann (1980).

We now outline what is contained in the various sections; at the end of each section there are some bibliographic notes to be found. The main results concerning the iterates of functions in $\mathcal{S}$ (Theorems 5.1 and 5.2) are stated in Section 5..It is possible for the reader to start at Section 5, and in order to make this easier we give an index of symbols at the end of the notes.

Section 2: Piecewise monotone functions Let $a, b \in \mathbb{R}$ with $a < b$ and let $f : [a,b] \longrightarrow [a,b]$ be continuous; f is called piecewise monotone if there exist $N \geq 0$ and $a = d_0 < d_1 < \cdots < d_N < d_{N+1} = b$ such that f is strictly monotone on each of the intervals $[d_k, d_{k+1}]$, $k = 0,\ldots,N$. Section 2 deals with some elementary properties of such functions. Our interest in this class of functions lies in the fact that if $f \in S$ then f^n is piecewise monotone for each $n \geqq 1$. Moreover, it is often just as easy to obtain results for piecewise monotone functions as it is for functions in S.

The section is mainly concerned with periodic points, and besides the usual classification of periodic points as being either stable, one-sided stable or unstable we also introduce the notion of a periodic point being "trapped": If x is periodic with period n then we say that x is trapped if there exist $y < x < z$ and $\delta > 0$ such that f^{2n} is increasing on the interval $[y-\delta, z+\delta]$ and $f^{2n}(y) \leqq y$, $f^{2n}(z) \geqq z$. The most important example of such a point is when a stable periodic point x is "trapped" between two unstable periodic points y and z:

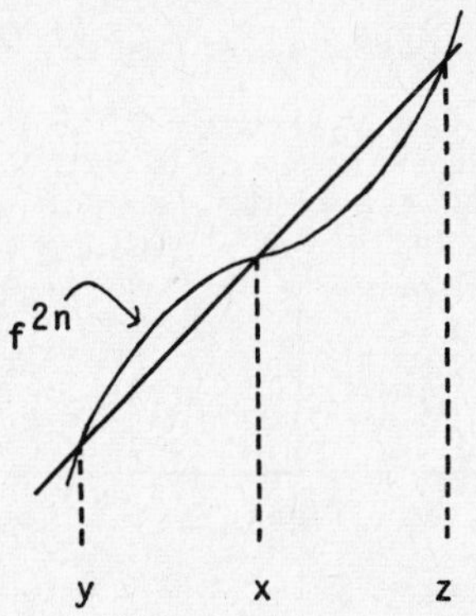

We will see that if a periodic point x is trapped then so are all the points in the periodic orbit $[x] = \{x, f(x), \ldots, f^{n-1}(x)\}$, and thus it also makes sense to talk about a periodic orbit being trapped.

The main result of the section implies that if $f \in S$ then f has at most one periodic orbit $[x] = \{x, f(x), \ldots, f^{n-1}(x)\}$ such that *(i)* $[x]$ is either stable or one-sided stable, *(ii)* $[x]$ is not trapped, and *(iii)* x is not a fixed point of f in $[0,\varphi)$ (where φ is the turning point of f). Moreover, if this orbit $[x]$ exists then for some $\delta > 0$ all the points in $(\varphi-\delta,\varphi)\cup(\varphi,\varphi+\delta)$ are attracted to $[x]$. This result is based on the proof of a theorem in Singer (1978). The existence or non-existence of the orbit $[x]$ will be important in the later sections for determining what kind of behaviour occurs for the iterates of f .

Section 3: Well-behaved piecewise monotone functions For a piecewise monotone function $f : [a,b] \longrightarrow [a,b]$ we let $A(f)$ denote the set of points in $[a,b]$ which are attracted to some periodic orbit of f ; it is easily seen that in fact

$$A(f) = \{ x \in [a,b] : \lim_{m\to\infty} f^{nm}(x) \text{ exists for some } n \geq 1 \},$$

and thus in some sense $A(f)$ consists of those points y in $[a,b]$ for which the orbit $\{f^n(y)\}_{n\geq 0}$ has a particularly simple behaviour. The aim of Section 3 is to find sufficient conditions under which $A(f)$ contains a dense open subset of $[a,b]$, i.e. under which a "typical" point in $[a,b]$ gets attracted to a periodic orbit. (See Remark 4 above.) We are also interested in knowing when $A'(f)$ contains a dense open subset of $[a,b]$, where $A'(f)$ is the set of points in $[a,b]$ which are attracted to either a stable or a one-sided stable periodic orbit.

In order to state what the main result of Section 3 says for functions in $\mathcal{S}$ let us fix $f \in \mathcal{S}$ with turning point φ and put

$$\gamma = \begin{cases} \text{the largest fixed point of } f \text{ in } [0,\varphi] & \text{if } f \text{ has a fixed point in } [0,\varphi] , \\ 0 & \text{otherwise.} \end{cases}$$

It is convenient to divide things into three cases:

I. $f(\varphi) \leqq \varphi$,

II. $f(\varphi) > \varphi$ and $f^2(\varphi) < \gamma$,

III. $f(\varphi) > \varphi$ and $f^2(\varphi) \geqq \gamma$.

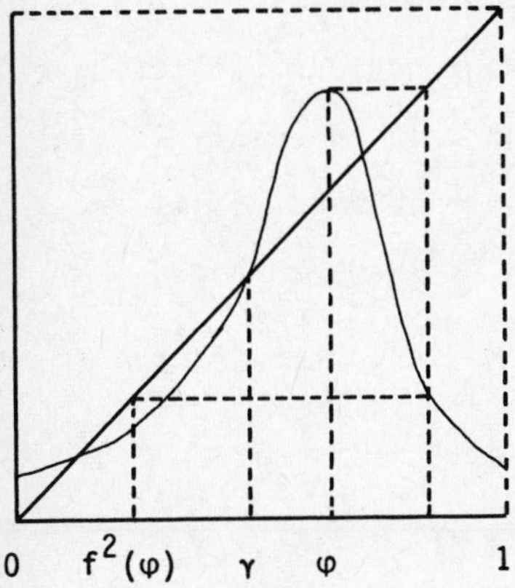

Case II

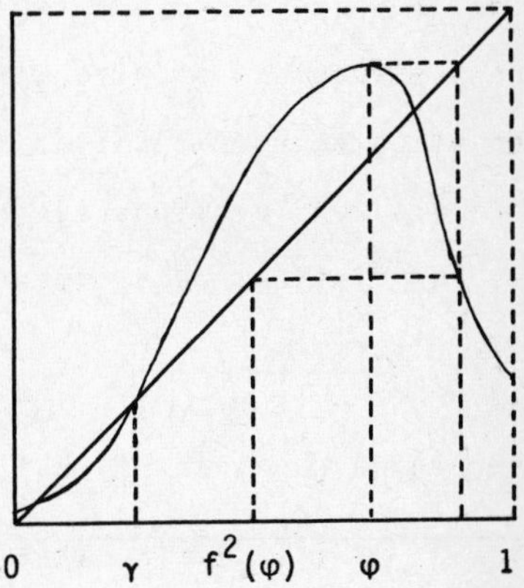

Case III

Case I is trivial and it is a simple matter to check that $A(f) = [0,1]$.

Case II: Here we will see that $A(f)$ contains a dense open subset of $[0,1]$ provided

(1.1) f has a continuous second derivative in $(0,\varphi) \cup (\varphi,1)$ and $f'(x) \neq 0$ for all $x \in (0,\varphi) \cup (\varphi,1)$.

Case III: This is the most interesting situation. We will show that $A(f)$ contains a dense open subset of $[0,1]$ provided (1.1) holds and one of the following three conditions is satisfied:

(1.2) φ is attracted to a stable periodic orbit $[y]$;

(1.3) φ is attracted to a one-sided stable periodic orbit $[y]$ but $f^k(\varphi) \neq y$ for all $k \geq 0$;

(1.4) the periodic orbit $[x]$ described in the main result of Section 2 exists.

The results stated above remain true for most functions in $\mathcal{S}$ when $A(f)$ is replaced by $A'(f)$.

Section 4: Property R and negative Schwarzian derivatives This section considers the subclass of piecewise monotone functions consisting of the functions which have what Collet and Eckmann call "property R". The importance of such functions is that all their trapped orbits are unstable. For distinct real numbers w , x , y and z put $R(w,x,y,z) = \dfrac{(z-w)(y-x)}{(z-y)(x-w)}$; if $f : [a,b] \to \mathbb{R}$ is continuous and strictly monotone then we say that f has property R if

$$R(f(x_1),f(x_2),f(x_3),f(x_4)) > R(x_1,x_2,x_3,x_4)$$

whenever $a < x_1 < x_2 < x_3 < x_4 < b$. A piecewise monotone function is said to have property R if the restriction of this function to any interval on which it is monotone has property R . The functions in the two families $\{f_\mu\}_{0<\mu\leq 4}$ and $\{g_r\}_{r>1}$ all have property R ; this is a consequence of the following fact which will be proved in Section 4: Let $f : [a,b] \to \mathbb{R}$ be continuous and have a third derivative in (a,b) ; suppose $f'(x) \neq 0$ for all $x \in (a,b)$. Then f has property R if and only if $(Sf)(x) < 0$ for all $x \in (a,b)$, where Sf , the Schwarzian derivative of f , is given by $(Sf)(x) = \left[\dfrac{f''(x)}{f'(x)}\right]' - \dfrac{1}{2}\left[\dfrac{f''(x)}{f'(x)}\right]^2$.

We will show that if a piecewise monotone function has property R then all its trapped orbits are unstable. This allows us to "improve" many of the results from Sections 2 and 3. Amongst other things we will prove the following:

Let $f \in \mathcal{S}$ have property R ; suppose (1.1) holds and that $f(z) > z$ for all $z \in (0,\varphi]$. Then f has at most one stable or one-sided stable periodic orbit. If this orbit exists then the set of points which get attracted to it contains a dense open subset of $[0,1]$; if it does not exist then $A(f)$ is countable.

Section 5: The iterates of functions in $\mathcal{S}$ This section gives the main result of these notes (Theorem 5.2), which classifies the types of behaviour that can be exhibited by the iterates of functions in $\mathcal{S}$. As a special case of this classification we obtain the following version of results due to Guckenheimer and Misiurewicz (Guckenheimer (1979), Misiurewicz (1980)): Let $f \in \mathcal{S}$ satisfy:

(1.5) f has property R ,

(1.6) f has a continuous second derivative in $[0,1]$, $f'(z) \neq 0$ for all $z \in [0,\varphi)\cup(\varphi,1]$ and $f''(\varphi) < 0$,

(1.7) $f(z) > z$ for all $z \in (0,\varphi]$.

Then exactly one of the following statements holds:

(1.8) f has a stable or a one-sided stable periodic orbit $[x]$ and the set of points which are attracted to $[x]$ contains a dense open subset of $[0,1]$.

(1.9) f has no stable or one-sided stable periodic orbits and f has sensitive dependence to initial conditions. (By the latter we will mean that $\Sigma_\varepsilon(f) = [0,1]$ for some $\varepsilon > 0$, where

$$\Sigma_\varepsilon(f) = \{\, x \in [0,1] : \text{for each } \delta > 0 \text{ there exists } n \geq 0 \text{ such that } |f^k((x-\delta,x+\delta))| \geq \varepsilon \text{ for all } k \geq n \,\},$$

and where $|J|$ denotes the length of the interval J .)

(1.10) f has no stable or one-sided stable periodic orbits and there exists a decreasing sequence $\{I^{(n)}\}_{n\geq 1}$ of closed subsets of $[0,1]$, each containing φ in its interior, such that $(\{I^{(n)}\}_{n\geq 1},f)$ is a proper infinite register shift; moreover, for each $n \geq 1$ $\{ x \in [0,1] : f^k(x) \in \mathrm{int}(I^{(n)})$ for some $k \geq 0 \}$ is a dense open subset of $[0,1]$.

(We say that $(\{I^{(n)}\}_{n\geq 1},f)$ is an infinite register shift if for each $n \geq 1$ there exists an integer $m_n \geq 2$ such that, putting $q_n = \prod_{k=1}^{n} m_k$, we have $I^{(n)}$ is the disjoint union of q_n non-trivial closed intervals $I_0^{(n)},\dots,I_{q_n-1}^{(n)}$ such that $f(I_{j-1}^{(n)}) = I_j^{(n)}$ for $j = 1,\dots,q_n-1$ and $f(I_{q_n-1}^{(n)}) = I_0^{(n)}$. We call an infinite register shift proper if $\bigcap_{n\geq 1} I^{(n)}$ is a nowhere dense subset of $[0,1]$.)

(1.8), (1.9) and (1.10) give a precise formulation of the three types of behaviour mentioned earlier, and the above result shows that if $f \in \mathcal{S}$ satisfies (1.5), (1.6) and (1.7) then these three types completely classify the possible behaviour which can be exhibited by the iterates of f. (In Section 10 we will show that each of the types actually occurs.)

Note that if (1.10) holds and we put $G = \bigcap_{n\geq 1} G_n$, where $G_n = \{ x \in [0,1] : f^k(x) \in \mathrm{int}(I^{(n)})$ for some $k \geq 0 \}$, then G is a residual subset of $[0,1]$ and the points in G are attracted to the Cantor-like set $I^{(\infty)} = \bigcap_{n\geq 1} I^{(n)}$: If $x \in G$ then for each $n \geq 1$ there exists $k \geq 0$ such that $f^j(x) \in I^{(n)}$ for all $j \geq k$. In particular $\lim_{n\to\infty} \min_{z\in I^{(\infty)}} |f^n(x)-z| = 0$ for each $x \in G$.

Section 6: Reductions and *Section 7: Getting rid of homtervals* These sections are concerned with the proofs of Theorems 5.1 and 5.2. The proof of Theorem 5.2 is based on a construction which involves what we call a reduction: Let $f \in S$, $0 \leq c < \varphi < d \leq 1$ and $m > 1$; we call $\Gamma = ([c,d], f^m)$ a reduction of $([0,1], f)$ if

(1.11) $f^m([c,d]) \subset [c,d]$,

(1.12) the intervals $(c,d), f((c,d)), \ldots, f^{m-1}((c,d))$ are disjoint.

Let $f \in S$ and $\Gamma = ([c,d], f^m)$ be a reduction of $([0,1], f)$; let g denote the restriction of f^m to $[c,d]$. It is not hard to see that φ is the only turning point of f^m in $[c,d]$ (see Lemma 6.1) and so g is either a scaled down or a scaled down and turned upside down version of a function from S .

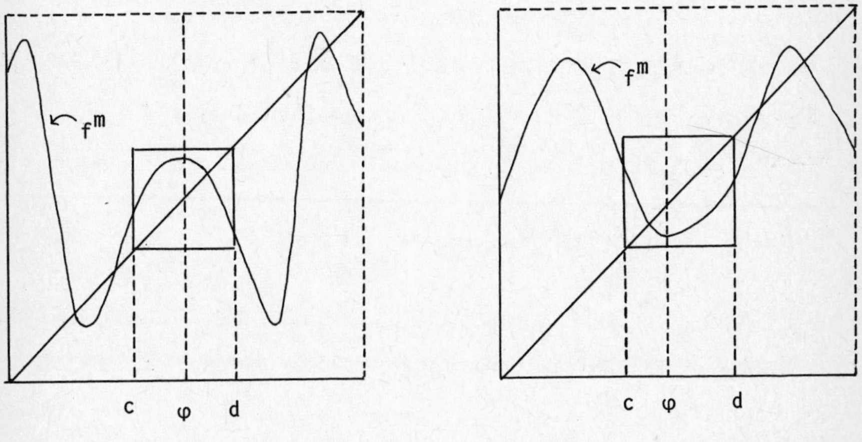

We can thus transform g into an element of S by a linear change of variables. Let us denote the element of S obtained in this way by $U_\Gamma f$, so $U_\Gamma f = \Psi_\Gamma \circ g \circ \Psi_\Gamma^{-1}$, where $\Psi_\Gamma : [c,d] \to [0,1]$ is given by

$$\Psi_\Gamma(x) = \begin{cases} \dfrac{x-c}{d-c} & \text{if } f^m \text{ is increasing on } [c,\varphi] , \\ \dfrac{d-x}{d-c} & \text{if } f^m \text{ is increasing on } [\varphi,d] . \end{cases}$$

Thus in fact we have

$$(U_\Gamma f)(x) = \begin{cases} \dfrac{1}{d-c}\Big(f^m((d-c)x+c) - c \Big) & \text{if } f^m \text{ is increasing on } [c,\varphi] , \\ \dfrac{1}{d-c}\Big(d - f^m(d-(d-c)x) \Big) & \text{if } f^m \text{ is increasing on } [\varphi,d] . \end{cases}$$

The idea behind the proof of Theorem 5.2 for the special case of functions satisfying (1.5), (1.6) and (1.7) is now roughly the following: Let $\mathcal{R} = \{ f \in \mathcal{S} : f$ satisfies (1.5) and (1.6) $\}$; if $f \in \mathcal{R}$ and Γ is a reduction of $([0,1],f)$ then it is easy to check that also $U_\Gamma f \in \mathcal{R}$. Put $\mathcal{R}_1 = \{ f \in \mathcal{R} : f(\varphi) \leqq \varphi \}$, $\mathcal{R}_2 = \{ f \in \mathcal{R} :$ (1.8) holds for $f \}$ and $\mathcal{R}_3 = \{ f \in \mathcal{R} :$ (1.9) holds for $f \}$. In Section 6 we will construct a subset $\mathcal{R}^*$ of $\mathcal{R}$ with $\mathcal{R}-\mathcal{R}^* \subset \mathcal{R}_1 \cup \mathcal{R}_3$ and for each $f \in \mathcal{R}^*$ a reduction $\Gamma(f)$ of $([0,1],f)$ such that

(1.13) if $U_{\Gamma(f)} f \in \mathcal{R}_1 \cup \mathcal{R}_2$ then $f \in \mathcal{R}_2$,

(1.14) if $U_{\Gamma(f)} f \in \mathcal{R}_3$ then $f \in \mathcal{R}_3$.

Let us define $Z : \mathcal{R} \longrightarrow \mathcal{R}$ by

$$Zf = \begin{cases} f & \text{if } f \notin \mathcal{R}^* , \\ U_{\Gamma(f)} f & \text{if } f \in \mathcal{R}^* , \end{cases}$$

and for $n \geq 1$ let $Z^n : \mathcal{R} \longrightarrow \mathcal{R}$ be given inductively by $Z^1 f = f$ and $Z^{n+1} f = Z(Z^n f)$. Now take $f \in \mathcal{S}$ satisfying (1.5), (1.6) and (1.7);

since $R-R^* \subset R_1 \cup R_3$ we have that either *(i)* $Z^n f \in R_1$ for some $n \geqq 1$, *(ii)* $Z^n f \in R_3$ for some $n \geqq 1$, or *(iii)* $Z^n f \in R^*$ for all $n \geqq 1$. If *(i)* holds then by (1.13) f satisfies (1.8), and if *(ii)* holds then by (1.14) f satisfies (1.9); in Section 6 we will show that if *(iii)* holds then f satisfies (1.10). This will give us that at least one of (1.8), (1.9) and (1.10) holds for f. The proof that f has at most one of these properties is straightforward.

Section 8: Kneading sequences Let

$$W = \{ \{\theta_n\}_{n\geqq 0} : \theta_n \in \{-1,0,1\} \text{ for each } n \geqq 0 \}$$

and for $f \in S$ define $k_f = \{k_f(n)\}_{n\geqq 0} \in W$ by

$$k_f(f) = \begin{cases} -1 & \text{if } f^n(\varphi) < \varphi , \\ 0 & \text{if } f^n(\varphi) = \varphi , \\ 1 & \text{if } f^n(\varphi) > \varphi . \end{cases}$$

k_f is called the kneading sequence of f and in this section we consider how much k_f tells us about a function $f \in S$. We will show that if $f \in S$ satisfies (1.5), (1.6) and (1.7) and if φ is not periodic then we can determine which one of (1.8), (1.9) and (1.10) occurs from just knowing k_f.

Section 9: An "almost all" version of Theorem 5.1 Let $f \in S$ satisfy (1.5), (1.6) and (1.7); we will prove that:

(1) If f has a stable periodic orbit $[x]$ then the set of points which are attracted to $[x]$ has Lebesgue measure one.

(2) If (1.10) holds and $(\{I^{(n)}\}_{n\geqq 1}, f)$ is the corresponding proper infinite register shift then for each $n \geqq 1$

$$\lambda(\{ x \in [0,1] : f^k(x) \in \mathrm{int}(I^{(n)}) \text{ for some } k \geq 0 \}) = 1$$

(where λ denotes Lebesgue measure).

Section 10: Occurrence of the different types of behaviour The main result of Section 5 shows that if $f \in S$ satisfies (1.5), (1.6) and (1.7) then exactly one of (1.8), (1.9) and (1.10) holds. In this section we will see that each of these three types actually occurs. In fact we will show that any "reasonable" one-parameter family of functions satisfying (1.5), (1.6) and (1.7) (such as $f_\mu(x) = \mu x(1-x)$, $2 < \mu \leqq 4$, $f_\mu(x) = \sin(\mu x)$, $\frac{\pi}{2} < \mu \leqq \pi$, and $f_\mu(x) = \mu x \exp(1-\mu x)$, $\mu > 1$) contains infinitely many functions of each type.

2. PIECEWISE MONOTONE FUNCTIONS

For $a, b \in \mathbb{R}$ with $a < b$ let $C([a,b])$ denote the set of continuous functions $f : [a,b] \rightarrow [a,b]$ which map the closed interval $[a,b]$ back into itself. If $f \in C([a,b])$ and $n \geq 0$ then f^n will denote the *n th. iterate* of f, i.e. $f^n \in C([a,b])$ is defined inductively by $f^0(x) = x$, $f^1(x) = f(x)$ and $f^n(x) = f(f^{n-1}(x))$. Let S consist of those functions $f \in C([0,1])$ for which there exists $\varphi \in (0,1)$ such that f is strictly increasing on $[0,\varphi]$ and is strictly decreasing on $[\varphi,1]$; these notes will be concerned with the iterates of functions in S. A function $f \in S$ might look something like the following picture:

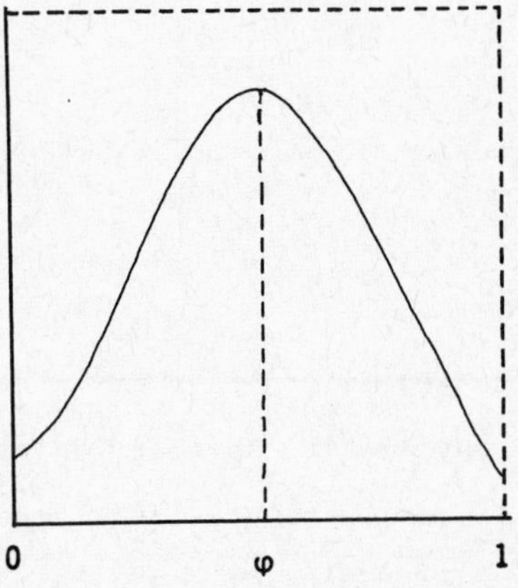

This section deals with some elementary properties of piecewise monotone functions; we are interested in this class of functions because it contains all the iterates of functions in S. We say that $f \in C([a,b])$ is *piecewise monotone* if there exist $N \geq 0$ and $a = d_0 < d_1 < \ldots < d_N < d_{N+1} = b$ such that f is strictly monotone on each of the intervals $[d_k,d_{k+1}]$, $k = 0,\ldots,N$. Let f be piecewise monotone and let us make the minimal choice for the d_k's , i.e. so that for $1 \leq k \leq N$ we have f is not monotone in any neighbourhood of d_k ; we then call $d_1,\ldots,d_N$ the *turning points* of f and the intervals $[d_k,d_{k+1}]$, $k = 0,\ldots,N$, the *laps* of f . Let $\mathcal{M}([a,b])$ denote the set of piecewise monotone functions in $C([a,b])$.

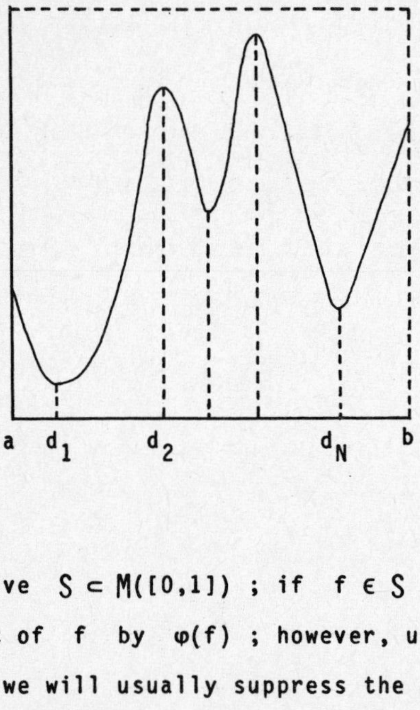

We clearly have $S \subset \mathcal{M}([0,1])$; if $f \in S$ then we denote the single turning point of f by $\varphi(f)$; however, unless there is any danger of confusion we will usually suppress the dependence on f and just write φ instead of $\varphi(f)$. $\mathcal{M}([a,b])$ is closed under composition: if $f, g \in \mathcal{M}([a,b])$ then also $f \circ g \in \mathcal{M}([a,b])$; thus in particular we have $f^n \in \mathcal{M}([0,1])$ for any $f \in S$, $n \geq 0$. For $f \in \mathcal{M}([a,b])$ let $T(f)$ denote the set of turning points of f ; the following fact will be needed repeatedly.

Proposition 2.1 Let $f \in \mathcal{M}([a,b])$ and $n \geq 1$; then

$$T(f^n) = \{ x \in (a,b) : f^k(x) \in T(f) \text{ for some } 0 \leq k < n \} .$$

In particular, for $f \in \mathcal{S}$ and $n \geq 1$ we have

$$T(f^n) = \{ x \in (0,1) : f^k(x) = \varphi \text{ for some } 0 \leq k < n \} .$$

Proof Suppose $x \in (a,b)$ is such that $f^k(x) \notin T(f)$ for all $0 \leq k < n$; then f is monotone in a neighbourhood of $f^k(x)$ for each $0 \leq k < n$ and thus f^n is monotone in some neighbourhood of x ; hence $x \notin T(f^n)$. Conversely, suppose $f^k(x) \in T(f)$ for some $0 \leq k < n$ and let $m = \min \{ k \geq 0 : f^k(x) \in T(f) \}$; then, since f is monotone in a neighbourhood of $f^k(x)$ for each $0 \leq k < m$, it is easy to see that $x \in T(f^{m+1})$. It is thus sufficient to show that $T(f^j) \subset T(f^{j+1})$ for each $j \geq 1$. But if $x \in T(f^j)$ and $\delta > 0$ is small enough then $f^j(x)$ is an end-point of the interval $J_\delta = f^j([x-\delta,x+\delta])$; thus we can choose $\delta > 0$ so that both $f^j(x)$ is an end-point of J_δ and f is monotone on J_δ , and this gives us that $x \in T(f^{j+1})$. ▫▫

Much of the analysis of the iterates of functions in $\mathcal{S}$ will have to do with periodic points. For $f \in C([a,b])$ and $n \geq 1$ let $\mathrm{Per}(n,f)$ denote the set of *periodic points* of f with *period* n , i.e.

$$\mathrm{Per}(n,f) = \{ x \in [a,b] : f^n(x) = x , f^k(x) \neq x \text{ for } k = 1,\dots,n-1 \} ;$$

in particular we have $\mathrm{Per}(1,f) = \mathrm{Fix}(f)$, the set of *fixed points* of f . Note that $\mathrm{Per}(n,f) \subset \mathrm{Fix}(f^n)$ for each $n \geq 1$. For $x \in \mathrm{Per}(n,f)$ we let $[x]$ denote the periodic orbit containing x , that is $[x] = \{x,f(x),\dots,f^{n-1}(x)\}$. We consider $[x]$ as a subset of $[a,b]$, and thus if $x, y \in \mathrm{Per}(n,f)$ then $[x] = [y]$ if and only if $y = f^k(x)$ for some $0 \leq k < n$. Let $P(f)$ denote the set of all periodic orbits of f .

We distinguish between three kinds of periodic points: stable, one-sided stable and unstable. First we consider the fixed points; $x \in Fix(f)$ is called *stable* if there exists a non-trivial interval $J \subset [a,b]$ with x in the interior of J such that $\lim_{m\to\infty} |f^m(J)| = 0$. (If $J \subset [a,b]$ is an interval then $|J|$ denotes the length of J ; we say J is *non-trivial* if $|J| > 0$. Note that if J is an interval then so is $f^m(J)$.) $x \in Fix(f)$ is called *one-sided stable* if it is not stable but there exists a non-trivial interval $J \subset [a,b]$ having x as an end-point such that $\lim_{m\to\infty} |f^m(J)| = 0$. If $x \in Fix(f)$ is neither stable nor one-sided stable then we call it *unstable*. Now let $x \in Per(n,f)$; we call x *stable* if it is a stable fixed point of f^n ; similarly we call x *one-sided stable* (resp. *unstable*) if it is a one-sided stable (resp. unstable) fixed point of f^n .

Remark: In the definition of a stable fixed point the interior of J means with respect to the relative topology on $[a,b]$; in particular this implies that if an end-point of the interval is a fixed point then it is either stable or unstable.

Suppose $f \in C([a,b])$ has a continuous derivative in $[a,b]$ (taking the one-sided derivatives at the end-points); let $x \in Fix(f)$. It is easily seen that x is stable if $|f'(x)| < 1$ and unstable if $|f'(x)| > 1$. The situation when $|f'(x)| = 1$ is more complicated; if $x \in (a,b)$ and $f'(x) = 1$ then x is stable if $f(x+\delta) < x+\delta$ and $f(x-\delta) > x-\delta$ for all small enough $\delta > 0$, it is one-sided stable if only one of these two conditions is satisfied, and it is unstable if neither are. If $x \in (a,b)$ and $f'(x) = -1$ then x is stable if $f^2(x+\delta) < x+\delta$ for all small enough $\delta > 0$, and it is unstable when this condition is not satisfied. There are also similar conditions for the case when $|f'(x)| = 1$ and x is either a or b ; we leave the reader to work out what these are. Consider now $x \in \mathrm{Per}(n,f)$; by the chain rule we have $(f^n)'(x) = \prod_{k=0}^{n-1} f'(f^k(x))$, thus x is stable if $\left| \prod_{k=0}^{n-1} f'(f^k(x)) \right| < 1$, unstable if $\left| \prod_{k=0}^{n-1} f'(f^k(x)) \right| > 1$, and what happens in the remaining case can also easily be determined.

Proposition 2.2 Let $f \in C([a,b])$ and $x \in \mathrm{Per}(n,f)$ be stable. Then $f^k(x)$ is also stable for each $0 \leq k < n$. Suppose that $f^{-1}(y)$ has no interior for each $y \in [a,b]$ and that $x \in \mathrm{Per}(n,f)$ is one-sided stable. Then $f^k(x)$ is one-sided stable for each $0 \leq k < n$.

Proof This is left as an easy exercise for the reader. Note that $f^{-1}(y)$ having no interior for each $y \in [a,b]$ implies that $f^m(J)$ is a non-trivial interval whenever J is. ▫▫

Let $f \in M([a,b])$ and $x \in \mathrm{Per}(n,f)$; $f^{-1}(y)$ is finite for each $y \in [a,b]$ and so it certainly has no interior, thus by Proposition 2.2 it makes sense to say that $[x]$ is either *stable*, *one-sided stable* or *unstable* depending on which of these types x itself is. For $f \in M([a,b])$ let $P_s(f)$ denote the set of elements in $P(f)$ which are either stable or one-sided stable and let $P_u(f)$ denote those which are unstable.

We now make a detailed analysis of the periodic points of piecewise monotone functions. If $f \in M([a,b])$ and x is a periodic point of f then we say that x is *critical* if $x \in [z]$ for some periodic $z \in T(f)$; we also apply this term to an orbit of critical points. Proposition 2.1 shows that if $x \in \mathrm{Per}(n,f)$ and $x \in (a,b)$ then x is critical if and only if $x \in T(f^n)$; the next result deals with the end-points of the interval.

Proposition 2.3 Let $f \in \mathcal{M}([a,b])$ and suppose one of the end-points of $[a,b]$ is periodic. Then this point is either critical, a fixed point of f, or has period 2, and this last alternative occurs if and only if $f(a) = b$ and $f(b) = a$.

Proof Without loss of generality assume it is b that is periodic. Suppose b is neither a fixed point nor critical. Since b is periodic but not a fixed point there exists $y \in [a,b)$ with $f(y) = b$, and clearly either $y = a$ or $y \in T(f)$. But y is also periodic and so we cannot have $y \in T(f)$ (since then b would be critical); thus $y = a$. The same argument now shows that $f(z) = a$ for some $z \in (a,b]$, and because $z \in T(f)$ would again imply that b is critical we must have $z = b$. ▫▫

Remarks: (2.1) Let $f \in \mathcal{M}([a,b])$ and suppose that $z \in T(f)$ is periodic; then it is easy to see that z cannot be one-sided stable. Thus if $[x]$ is critical then it is either stable or unstable. Moreover, if f has a continuous derivative in $[a,b]$ then any critical orbit must be stable. (This follows because for each $z \in T(f)$ we have $f'(z) = 0$ and thus also $(f^n)'(z) = \prod_{k=0}^{n-1} f'(f^k(z)) = 0$.)

(2.2) Let $f \in \mathcal{S}$; in this case $f(0) = 1$ is not possible and so if an end-point of $[0,1]$ is periodic then it must be either critical or a fixed point. We cannot have $f(1) = 1$ and thus 1 is periodic if and only if it is critical; it is clear that this happens if and only if φ is periodic and $f(\varphi) = 1$. Similarly, if 0 is not a fixed point then it is periodic if and only if φ is periodic, $f(\varphi) = 1$ and $f(1) = 0$.

(2.3) Let $f \in \mathcal{M}([a,b])$ and suppose $x \in Per(n,f)$ is not critical; then f^n is monotone in a neighbourhood of x and therefore f^{2n} is increasing in some neighbourhood of x .

We now make a classification of the non-critical periodic points. Let $f \in \mathcal{M}([a,b])$ and $x \in Per(n,f)$; we say that x is *trapped* if there exist $y < x < z$ and $\delta > 0$ such that f^{2n} is increasing on the interval $[y-\delta,z+\delta]$ and $f^{2n}(y) \leq y$, $f^{2n}(z) \geq z$.

Proposition 2.4 Let $f \in \mathcal{M}([a,b])$ and let $x \in (a,b)$ be a non-critical unstable periodic point of f ; then x is trapped.

Proof Let $x \in (a,b)$ be a non-critical periodic point with period n ; let $[u,v]$ be the largest interval containing x on which f^{2n} is increasing. By (2.3) we thus have $u < x < v$. Now suppose x is not trapped; then either $f^{2n}(z) > z$ for all $z \in (u,x)$ or $f^{2n}(z) < z$ for all $z \in (x,v)$; without loss of generality we can assume that the former holds. Choose $\bar{u} \in (u,x)$ and let $J = [\bar{u},x]$; J is thus a non-trivial interval having x as an end-point. But f^k is monotone on J for each $k \geq 0$ and so the interval $f^k(J)$ has $f^k(\bar{u})$ and $f^k(x)$ as its end-points; hence $\lim_{m\to\infty} |f^{nm}(J)| = \lim_{m\to\infty} |f^{nm}(\bar{u})-x| = 0$, because $\lim_{m\to\infty} f^{2nm}(\bar{u}) = x$ and also $\lim_{m\to\infty} f^{(2m+1)n}(\bar{u}) = f^n(x) = x$. Therefore x is at least one-sided stable. ▫▫

Proposition 2.4 shows that most unstable periodic points are trapped; however we will be more interested in the trapped periodic points that are either stable or one-sided stable. An example of such a point is when a stable periodic point x is "trapped" between two unstable periodic points y and z :

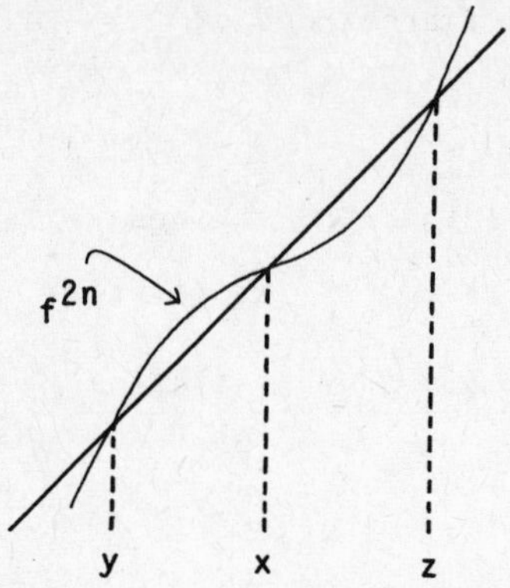

Note: In Section 4 we will study a class of functions for which all trapped periodic points are unstable. For such a function the above picture is then not possible.

Let $f \in \mathcal{M}([a,b])$ and let x be periodic; we say that x is *free* if it is neither critical nor trapped and if $[x] \subset (a,b)$. If a periodic point x is neither critical, trapped nor free then by Proposition 2.3 we have that x is an end-point of $[a,b]$ and is either a fixed point of f or has period 2 , this latter occurring if and only if $f(a) = b$ and $f(b) = a$.

Proposition 2.5 Let $f \in \mathcal{M}([a,b])$ and $x \in Per(n,f)$; then x is free (resp. trapped) if and only if $f^k(x)$ is free (resp. trapped) for each $0 \leq k < n$.

Proof Let $x \in Per(n,f)$ be trapped and $0 \leq k < n$. Now neither x nor $f^k(x)$ is critical or an end-point of $[a,b]$, and so if we let $[u,v]$ (resp. $[\bar{u},\bar{v}]$) be the largest interval containing x (resp.

$f^k(x)$) on which f^{2n} is increasing then $u < x < v$ and also $\bar{u} < f^k(x) < \bar{v}$. Since x is trapped there exist $u < y < x < z < v$ with $f^{2n}(y) \leq y$ and $f^{2n}(z) \geq z$, and it is clearly possible to choose y and z so that $f^{2n}(y) > u$ and $f^{2n}(z) < v$. By Proposition 2.1 f^k is monotone on $[u,v]$ and so we can define

$$\bar{y} = \begin{cases} f^k(y) & \text{if } f^k \text{ is increasing on } [u,v] , \\ f^k(z) & \text{otherwise,} \end{cases}$$

and let $\bar{z}$ be the other of the two points $f^k(y)$ and $f^k(z)$. Thus $\bar{y} < f^k(x) < \bar{z}$ and $f^{2n}(\bar{y}) \leq \bar{y}$, $f^{2n}(\bar{z}) \geq \bar{z}$. (For example, if f^k is increasing on $[u,v]$ then $f^{2n}(\bar{y}) = f^{2n}(f^k(y)) = f^k(f^{2n}(y))$ and $f^k(f^{2n}(y)) \leq f^k(y) = \bar{y}$; and the other cases are very similar.) Furthermore, it is easy to check that $\bar{u} < \bar{y}$ and $\bar{z} < \bar{v}$, and therefore $f^k(x)$ is trapped. Conversely, if $f^k(x)$ is trapped then the above proof also shows that x is trapped because $x = f^{n-k}(f^k(x))$. The other half of the result is now a direct consequence of the first half. ◻◻

By Proposition 2.5 it makes sense to say that the periodic orbit $[x]$ is either *free* or *trapped* if x itself is of the corresponding type. From Proposition 2.4 we know that if $[x]$ is free then $[x] \in P_s(f)$.

We now come to the main result of this section, which says that if $f \in \mathcal{M}([a,b])$ has N turning points then

$$|\{ [x] \in P_s(f) : [x] \text{ is not trapped } \}| \leq N+\Delta ,$$

where $\Delta = \begin{cases} 1 & \text{if } N \text{ is odd,} \\ 2 & \text{if } N \text{ is even.} \end{cases}$ In particular, if $f \in \mathcal{S}$ then $P_s(f)$

contains at most 2 elements which are not trapped; we will in fact see that in this case $\{ [x] \in P_s(f) : [x]$ is not trapped $\}$ contains at most one element which is not a fixed point in the interval $[0,\varphi)$. As we have already mentioned, in Section 4 we will study a class of functions for which all trapped periodic points are unstable. For such a function $f \in M([a,b])$ with N turning points we thus have $|P_s(f)| \leq N+\Delta$.

The result which we have just stated will follow from a result about free orbits; this says that if $f \in M([a,b])$ then, except possibly for two exceptional orbits, to each free orbit $[x]$ there exists a $w \in T(f)$ such that all points near w get attracted to the orbit $[x]$. To make this precise we need a couple of definitions. If $f \in C([a,b])$ and $x \in Per(n,f)$ then we let $A([x],f)$ denote the set of points in $[a,b]$ which are attracted to the orbit $[x]$, i.e.

$$A([x],f) = \{ y \in [a,b] : \lim_{m\to\infty} f^{nm}(y) = f^k(x) \text{ for some } 0 \leq k < n \} .$$

Note that if $[x] \neq [z]$ then $A([x],f)$ and $A([z],f)$ are disjoint.

Let $f \in M([a,b])$ and let $[x]$ be a non-critical periodic orbit with $[x] \subset (a,b)$; we say that $[x]$ is *exceptional* if one of the following holds:

(2.4) $x \in Fix(f)$, f is increasing on $[x,b]$ and $f(z) < z$ for all $z \in (x,b)$,

(2.5) $x \in Fix(f)$, f is increasing on $[a,x]$ and $f(z) > z$ for all $z \in (a,x)$,

(2.6) $x \in Per(2,f)$, and if x is chosen so that $x < f(x)$ then

f is decreasing on $[a,x]$ and $[f(x),b]$, and $f^2(z) > z$ for all $z \in (a,x)$.

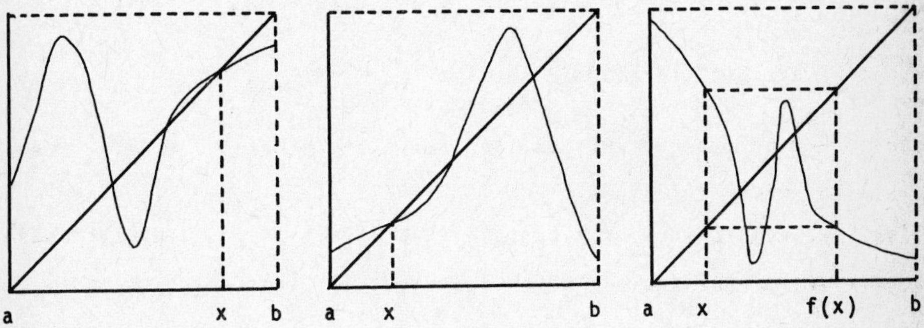

(2.4) (2.5) (2.6)

It is easy to see that if $[x]$ is exceptional then it is free. Each of (2.4), (2.5) and (2.6) can be satisfied by at most one orbit, and if there is an orbit satisfying (2.6) then there are none satisfying (2.4) and (2.5). Thus there are at most two exceptional orbits. If $f \in S$ then only (2.5) can occur: we say that $x \in (0,\varphi)$ is an *initial fixed point* if $f(x) = x$ and $f(z) > z$ for all $z \in (0,x)$; clearly (2.5) holds if and only if x is an initial fixed point.

Proposition 2.6 Let $f \in M([a,b])$ and let $[x]$ be a free, non-exceptional periodic orbit. Then there exists $\delta > 0$ and $w \in T(f)$ such that $(w-\delta,w)\cup(w,w+\delta) \subset A([x],f)$. (In fact we also have $w \in A([x],f)$ unless w is an unstable periodic point.) In particular, if $f \in S$ and $[x]$ is a free orbit which is not an initial fixed point then for some $\delta > 0$ we have $(\varphi-\delta,\varphi)\cup(\varphi,\varphi+\delta) \subset A([x],f)$.

Proof Let $[x]$ be free; suppose x has period n and let $[u,v]$ be the largest interval containing x on which f^{2n} is increasing. Then $u < x < v$ and either $f^{2n}(z) < z$ for all $z \in (x,v)$ or $f^{2n}(z) > z$ for all $z \in (u,x)$. Assume first that $f^{2n}(z) < z$ for all $z \in (x,v)$; then $\lim_{m\to\infty} f^{2nm}(z) = x$ for all $z \in (x,v)$, and thus $\lim_{m\to\infty} f^{nm}(z) = x$ for all $z \in (x,v)$ (because also $\lim_{m\to\infty} f^{n(2m+1)}(z) = f^n(x) = x$). Now suppose $v \neq b$; then by Proposition 2.1 there exists $w \in T(f)$ and $0 \leq k < 2n$ with $f^k(v) = w$. Let $J = f^k((x,v))$, so J is a non-trivial interval with $w \in \bar{J}$ (the closure of J) and we have $J \subset A([x],f)$, because if $y \in J$ then $y = f^k(z)$ for some $z \in (x,v)$ and hence $\lim_{m\to\infty} f^{nm}(y) = f^k(x)$. Therefore for some $\varepsilon > 0$ we have either $(w-\varepsilon,w) \subset A([x],f)$ or $(w,w+\varepsilon) \subset A([x],f)$, and this implies that $(w-\delta,w)\cup(w,w+\delta) \subset A([x],f)$ for some $\delta > 0$ (since if $\varepsilon > 0$ is small enough then $f((w-\varepsilon,w)) = f((w,w+\bar{\varepsilon}))$, where $w+\bar{\varepsilon}$ is the smallest point to the right of w with $f(w-\varepsilon) = f(w+\bar{\varepsilon})$). Note also that if w is not a fixed point of f^{2n} then we have $\lim_{m\to\infty} f^{nm}(w) = f^k(x)$ and thus $w \in A([x],f)$. If w is a fixed point of f^{2n} then it must be an unstable periodic point of f. (Suppose it was not unstable; then we would have $(w-\varepsilon,w+\varepsilon) \subset A([w],f)$ for some $\varepsilon > 0$ and thus $A([x],f)$ and $A([w],f)$ would not be disjoint; but this would imply that $[x] = [w]$, and this cannot happen because x is not critical.) The same proof works if $f^{2n}(z) > z$ for all $z \in (u,x)$ and $u \neq a$; we are thus left with the two cases when *(i)* f^{2n} is increasing on $[x,b]$ and $f^{2n}(z) < z$ for all $z \in (x,b)$, and *(ii)* f^{2n} is increasing on $[a,x]$ and $f^{2n}(z) > z$ for all $z \in (a,x)$. The analysis of these two cases involves the exceptional orbits, and since this is somewhat involved we leave the remainder of the proof to the end of the section. ▫▫

Remark: Suppose $f \in \mathcal{M}([a,b])$ has a continuous derivative in $[a,b]$; then (as noted in (2.1)) all critical points are stable. Thus in this case Proposition 2.6 gives us that if $[x]$ is a free, non-exceptional periodic orbit then there exists $\delta > 0$ and $w \in T(f)$ such that $(w-\delta, w+\delta) \subset A([x],f)$.

We now give two corollaries of Proposition 2.6; the first is the already mentioned main result of this section.

Theorem 2.1 (1) Let $f \in \mathcal{M}([a,b])$ have N turning points; then

$$|\{ [x] \in P_s(f) : [x] \text{ is not trapped } \}| \leq N+\Delta ,$$

where $\Delta = \begin{cases} 1 & \text{if } N \text{ is odd,} \\ 2 & \text{if } N \text{ is even.} \end{cases}$

(2) Let $f \in S$; then $\{ [x] \in P_s(f) : [x]$ is not trapped $\}$ contains at most two elements, and contains at most one element which is not a fixed point in the interval $[0,\varphi)$.

Proof (1): If $[x] \in P_s(f)$ is not trapped then one of the following holds:
(i) $[x]$ is critical, *(ii)* $[x]$ is free and non-exceptional,
(iii) $[x]$ is exceptional, *(iv)* $x = a$ is a fixed point of f ,
(v) $x = b$ is a fixed point of f , *(vi)* $[x] = \{a,b\}$ with $f(a) = b$ and $f(b) = a$.
But if $[x] \in P_s(f)$ is critical then $(x-\delta, x+\delta) \subset A([x],f)$ for some $\delta > 0$, and thus by Proposition 2.6 we have

$$|\{ [x] \in P_s(f) : \text{either } (i) \text{ or } (ii) \text{ holds } \}| \leq N .$$

Let $\Delta(f) = |\{ [x] \in P_s(f) : \text{either } (iii), (iv), (v) \text{ or } (vi) \text{ holds } \}|$; it is therefore enough to show that $\Delta(f) \leq \Delta$. If an exceptional orbit satisfying (2.4) (resp. (2.5)) exists then b (resp. a) cannot be a stable fixed point. Similarly, if an exceptional orbit satisfying (2.6) exists and $f(b) = a$, $f(a) = b$ then the orbit $\{a,b\}$ is unstable. From this it easily follows that $\Delta(f) \leq 2$. Furthermore, it is also easy to see that if $\Delta(f) = 2$ then f is increasing in some neighbourhood of a and in some neighbourhood of b , and this can only happen if N is even.

(2): This is a special case of (1). ⊞

The second application of Proposition 2.6 is to give examples of functions $f \in S$ for which $\{ [x] \in P_s(f) : [x] \text{ is not trapped } \}$ is either empty or contains at most a fixed point in the interval $[0,\varphi)$.

Theorem 2.2 (1) Let $f \in S$ with $f(y) > y$ for all $y \in (0,\varphi]$; suppose φ is not periodic but that $\varphi \in A([z],f)$ for some $[z] \in P_u(f)$. Then all the elements of $P_s(f)$ are trapped.

(2) Let $f \in S$ and suppose $f^2(\varphi) \leq \gamma$ for some fixed point $\gamma \in [0,\varphi)$. Then $\{ [x] \in P_s(f) : [x] \text{ is not trapped } \}$ consists of at most a fixed point in $[0,\varphi)$.

Note: If $f(\varphi) \leq \varphi$ then $f^n(\varphi) > \gamma$ for all $n \geq 1$ for any fixed point $\gamma \in [0,\varphi)$. Thus in (2) we must have $f(\varphi) > \varphi$. (See the picture on the next page.)

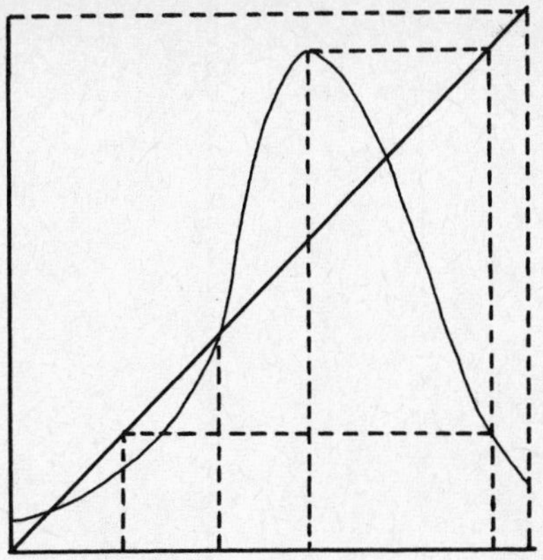

Proof (1): $f(y) > y$ for all $y \in (0,\varphi]$ implies that f has no initial fixed point and that 0 cannot be a stable fixed point; also f has no critical orbit because φ is not periodic. Therefore $\{ [x] \in P_s(f) : [x]$ is not trapped $\}$ contains at most a free, non-exceptional orbit $[v]$. But if such a $[v]$ exists then by Proposition 2.6 we would have $\varphi \in A([v],f)$ (since φ is not an unstable periodic point), and this would imply that $[v] = [z]$. This cannot happen and so $\{ [x] \in P_s(f) : [x]$ is not trapped $\}$ is empty.

(2): We have $f^n(\varphi) \leq \gamma < \varphi$ for all $n \geq 2$ and so φ is not periodic; thus f has no critical orbit. Suppose f has a free, non-exceptional orbit $[v]$, then by Proposition 2.6 we have $\varphi \in A([v],f)$; but $f^n(\varphi) \leq \gamma$ for all $n \geq 2$ also implies that $\lim_{m\to\infty} f^n(\varphi)$ exists, and thus v must be a fixed point of f in $[0,\gamma]$. However, if such a $[v]$ is free then it is an initial fixed point. (Let $[0,u]$ be the largest interval containing v on which f^2 is increasing; $f^2(z) < z$ for all $z \in (v,u)$ is not possible because $f(\varphi) > \varphi$.) Therefore no free, non-exceptional orbit exists and hence $\{ [x] \in P_s(f) : [x]$ is not trapped $\}$ consists of at most a fixed point in $[0,\varphi)$ (i.e. either an initial fixed point or 0 as a stable fixed point). □□

The simplest case of a function $f \in S$ satisfying the hypotheses of Theorem 2.2(1) is when we have $f(y) > y$ for all $y \in (0,\varphi]$, $f(0) = f(1) = 0$ and $f(\varphi) = 1$.

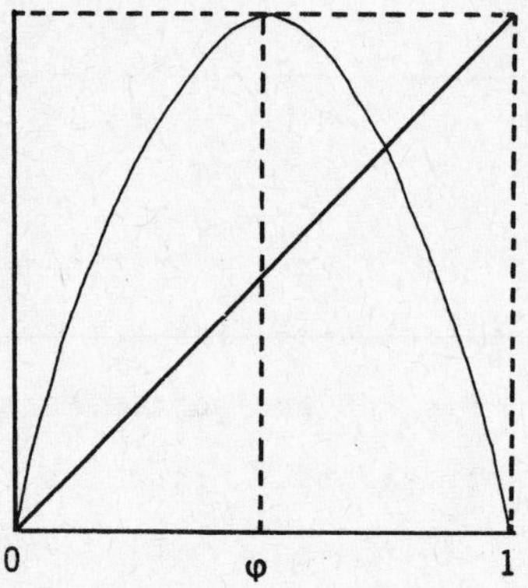

Here φ is attracted to the unstable fixed point 0. Other examples occur when $f(y) > y$ for all $y \in (0,\varphi]$ and $f^n(\varphi) = \beta$ for some $n \geq 1$, where β is the unique fixed point of f in $(\varphi,1)$. φ is then attracted to β, and so if β is unstable then the hypotheses are again satisfied. (See the picture on the next page.)

Remark: In the above examples $\varphi \in A([z],f)$ followed from having $f^k(\varphi) = z$ for some $k \geq 0$. For unstable orbits $[z]$ this is in fact the only possible way of being attracted to $[z]$: it is not difficult to show that if $f \in M([a,b])$ and $[z] \in P_u(f)$ then

$$A([z],f) = \{ y \in [a,b] : f^k(y) = z \text{ for some } k \geq 0 \}.$$

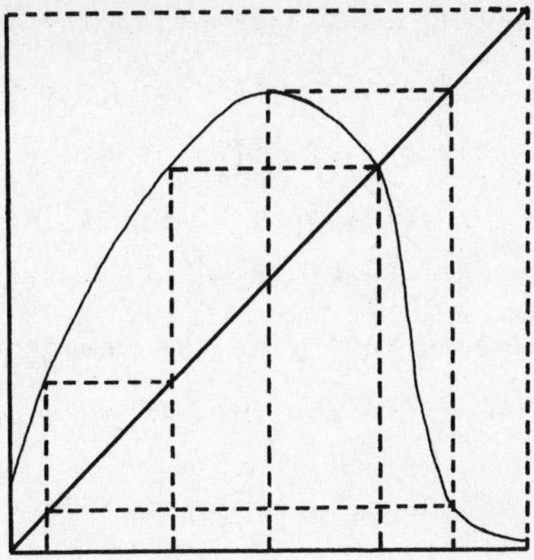

Theorems 2.1 and 2.2 can of course be best applied to functions for which all trapped orbits are unstable; we deal with this situation in Section 4.

If $f \in \mathcal{M}([a,b])$ then Theorem 2.1 gives us an upper bound on the number of elements in the set

$$P_s^*(f) = \{ [x] \in P_s(f) : [x] \text{ is not trapped } \} .$$

We now present a "practical" method for working out how large $P_s^*(f)$ is for a given function $f \in \mathcal{M}([a,b])$. Consider $f \in \mathcal{M}([a,b])$ to be fixed, and for simplicity we will assume in what follows that any critical orbit of f is stable. (As we have already remarked in (2.1), this holds, for example, if f has a continuous derivative in $[a,b]$.) Let $[x] \in P_s^*(f)$; then $[x]$ belongs to one of the six cases (*(i)* to *(vi)*) listed in the proof of Theorem 2.1. Now if either *(i)* or *(ii)* holds for $[x]$ then there exists $w \in T(f)$ with $w \in A([x],f)$. (If $[x]$ is critical then $w \in [x]$ for some $w \in T(f)$, and so we trivially have $w \in A([x],f)$; if $[x]$ is free and non-exceptional then

Proposition 2.6, together with the assumption that critical points are stable, implies that $w \in A([x],f)$ for some $w \in T(f)$.) This suggests that we should take each element of $T(f)$ and see if it is attracted to some periodic orbit; the periodic orbits obtained in this way will then include all elements of $P_s^*(f)$ for which either *(i)* or *(ii)* holds. It is then a simple matter to determine the remaining elements of $P_s^*(f)$ (i.e. those for which either *(iii)*, *(iv)*, *(v)* or *(vi)* holds).

The problem with this procedure is that it is possible for an element of $T(f)$ to be attracted to a trapped orbit. We can thus end up with too many orbits and must then decide which of them are in $P_s^*(f)$. However, there is a modified procedure for which this problem does not arise, and to explain how this works we need a couple of definitions. Suppose f has turning points $d_1 < d_2 < \cdots < d_N$; put $d_0 = a$ and $d_{N+1} = b$, and for $0 \leq k \leq N$ let $I_k = [d_k, d_{k+1}]$; also let $\Lambda = \{ \{\sigma_m\}_{m \geq 0} : \sigma_m \in \{0,1,\ldots,N\}$ for each $m \geq 0 \}$. We say that $\sigma = \{\sigma_m\}_{m \geq 0} \in \Lambda$ is a *code* for $x \in [a,b]$ if $f^m(x) \in I_{\sigma_m}$ for all $m \geq 0$. Clearly each point in $[a,b]$ has at least one code and the points in

$$G(f) = \{ x \in [a,b] : f^m(x) \notin T(f) \text{ for all } m \geq 0 \}$$

have exactly one code; $G(f)$ contains "most" of the points in $[a,b]$ since $[a,b] - G(f)$ is countable. We call $w \in T(f)$ *regular* if $f(w) \in G(f)$ and if one of the two possible codes for w is periodic. ($\sigma = \{\sigma_m\}_{m \geq 0} \in \Lambda$ is called *periodic* if there exists $n \geq 1$ such that $\sigma_{m+n} = \sigma_m$ for all $m \geq 0$; the smallest such $n \geq 1$ is called the *period* of σ . Note that any $v \in T(f)$ with $f(v) \in G(f)$ does have exactly two codes $\{\sigma_m\}_{m \geq 0}$ and $\{\bar{\sigma}_m\}_{m \geq 0}$ and these are such that

$\sigma_m = \bar{\sigma}_m$ for all $m \geq 1$ and $|\sigma_0 - \bar{\sigma}_0| = 1$.)

Proposition 2.7 If $w \in T(f)$ is regular then there exists $[x] \in P_s^*(f)$ such that $w \in A([x],f)$. Conversely, if $[x] \in P_s^*(f)$ is free and non-exceptional then there exists a regular $w \in T(f)$ with $w \in A([x],f)$.

Proof Let $w \in T(f)$ be regular and let $\sigma = \{\sigma_m\}_{m \geq 0}$ be the periodic code for w ; suppose σ has period n . Let J be the set of all points in $[a,b]$ which have σ as a code; then it is easily seen that J is a closed interval, f^m is monotone on J for each $m \geq 1$ and $f^n(J) \subset J$. Also J is non-trivial (since $w \in J$ and w is not periodic) and w is an end-point of J ; without loss of generality we can assume it is the right-hand end-point. Now f^{2n} must be increasing on J and $f^{2n}(J) \subset J$; moreover, $f^{2n}(w) < w$ (since w is not periodic) and thus $\lim_{m\to\infty} f^{2nm}(w) = x$, where x is the largest fixed point of f^{2n} in J . Therefore $w \in A([x],f)$ and it is easy to check that $[x] \in P_s^*(f)$. Conversely, let $[x] \in P_s^*(f)$ be free and non-exceptional and suppose $x \in Per(n,f)$. For $0 \leq k < n$ let $[u_k,v_k]$ be the largest interval containing $f^k(x)$ on which f^{2n} is increasing; thus $u_k < f^k(x) < v_k$. The remainder of the proof of Proposition 2.6 (at the end of the section) will show that for some $0 \leq k < n$ we have either $u_k \neq a$ and $f^{2n}(z) > z$ for all $z \in (u_k,f^k(x))$ or $v_k \neq b$ and $f^{2n}(z) < z$ for all $z \in (f^k(x),v_k)$; without loss of generality suppose it is the latter. By Proposition 2.1 there exists $0 \leq j < 2n$ and $w \in T(f)$ with $f^j(v_k) = w$; thus in fact we have $f^{2n}(z) < z$ for all $z \in (f^k(x),v_k]$ (since $f^{2n}(v_k) = v_k$ would imply that v_k is an unstable periodic point and hence $w \in T(f)$ would also be an unstable periodic point; by hypothesis this cannot happen). Therefore

$$\lim_{m\to\infty} f^{2nm}(w) = \lim_{m\to\infty} f^{2nm}(f^j(v_k)) = f^{k+j}(x) ,$$

and so $w \in A([x],f)$. It is now a simple matter to show that w is regular. (We have $I = f^j((f^k(x),v_k)) \subset G(f)$ and all the points in I have the same code σ ; this code is periodic and is one of the two codes for w .) ◻◻

Our modified procedure for finding the elements of $P_s^*(f)$ comes directly from Proposition 2.7:

Take $w \in T(f)$ and check first whether it is periodic; if it is then $[w] \in P_s^*(f)$. If w is not periodic then check whether it is regular; if it is then it is attracted to an element of $P_s^*(f)$. Apply this procedure to all the points in $T(f)$.

The elements of $P_s^*(f)$ obtained in this way will then include all the members of $P_s^*(f)$ for which either *(i)* or *(ii)* holds. Any remaining elements of $P_s^*(f)$ can then be easily found (and we leave the reader to work out how this should be done in practice).

Let us consider the above procedure for a function $f \in S$ (and here we know that $|P_s^*(f)| \le 2$). $P_s^*(f)$ will contain an element which is not a fixed point in $[0,\varphi)$ if and only if φ is either periodic or regular (and note in this case that if φ is not periodic then we automatically have $f(\varphi) \in G(f)$). $P_s^*(f)$ will contain an initial fixed point if and only if for all small enough $\delta > 0$ the sequence $\{f^m(\delta)\}_{m\ge 0}$ is strictly increasing. (If $f(0) > 0$ then this can be simplified to:

if and only if the sequence $\{f^m(0)\}_{m\geq 0}$ is strictly increasing.)

We leave it as an exercise for the reader to work out how the above procedure must be modified in order to handle the case when unstable critical orbits can occur.

Proof of Proposition 2.6 (continued) Let $x \in \mathrm{Per}(n,f)$ be free; we must deal with the two cases when *(i)* f^{2n} is increasing on $[x,b]$ and $f^{2n}(z) < z$ for all $z \in (x,b)$, and *(ii)* f^{2n} is increasing on $[a,x]$ and $f^{2n}(z) > z$ for all $z \in (a,x)$. In fact we need only deal with the more special situation when for each $0 \leq k < n$ we have either *(i)* f^{2n} is increasing on $[f^k(x),b]$ and $f^{2n}(z) < z$ for all $z \in (f^k(x),b)$, or *(ii)* f^{2n} is increasing on $[a,f^k(x)]$ and $f^{2n}(z) > z$ for all $z \in (a,f^k(x))$. (If neither *(i)* nor *(ii)* holds for some $0 \leq j < n$ then, since $f^j(x)$ is free, we can apply the part of the proof already given to $f^j(x)$.) But if *(i)* or *(ii)* holds for each $0 \leq k < n$ then clearly n is either 1 or 2. Suppose first that $n = 1$, that f^2 is increasing on $[x,b]$ and $f^2(z) < z$ for all $z \in (x,b)$. If f is increasing on $[x,b]$ then it is easy to see that (2.4) holds. On the other hand, if f is decreasing on $[x,b]$ then we must have $f^2(z) > z$ for all $z \in (u,x)$, where u is the largest turning point of f^2 to the left of x; such a case is covered by the first part of the proof. The other possibility with $n = 1$ (i.e. when f^2 is increasing on $[a,x]$ and $f^2(z) > z$ for all $z \in (a,x)$) can be handled in the same way. Suppose then that $n = 2$; we choose x so that $x < f(x)$ and thus we have f^4 is increasing on $[a,x]$ and $f^4(z) > z$ for all $z \in (a,x)$. If f^2 is decreasing on $[a,x]$ then, as above, we would have $f^4(z) < z$ for all $z \in (x,v)$, where v is the smallest turning point of f^4 to the right of x; this case is again covered by the first part of the proof. We can therefore assume that f^2 is increasing on $[a,x]$, and it follows that we have $f^2(z) > z$ for all $z \in (a,x)$. If f is decreasing on $[a,x]$ then it is also decreasing on $[f(x),b]$, and thus (2.6) holds. This leaves the case when f is increasing on $[a,x]$; here it is not difficult to show that $f^2(z) > z$ for all $z \in (\overline{u},f(x))$, where $\overline{u}$ is the largest turning point of f^2 to the left of $f(x)$, and once more this is covered by the first part of the proof. Finally, note that if x is also non-exceptional then the above proof gives us the fact which we have used in the proof of Proposition 2.7, namely, for some $0 \leq k < n$ either $u_k \neq a$ and $f^{2n}(z) > z$ for all $z \in (u_k,f^k(x))$ or $v_k \neq b$ and $f^{2n}(z) < z$ for all $z \in (f^k(x),v_k)$, where $[u_k,v_k]$ is the largest interval containing $f^k(x)$ on which f^{2n} is increasing. ⊞

Notes: Proposition 2.6 is based on the proof of Theorem 2.7 in Singer (1978). Proposition 2.7 can be found in Guckenheimer (1979); the "codes" used in this result will be studied in much more detail later in these notes, in particular in Section 8. Some general references concerning maps on an interval, which should give the reader some idea on how the interest in this subject has developed, are:

Ulam and von Neumann (1947), Parry (1964), Šarkovskii (1964), Stein and Ulam (1964), Metropolis, Stein and Stein (1973), Li and Yorke (1975), May (1975), (1976), Bowen and Franks (1976), May and Oster (1976), Oster, Ipaktchi and Rocklin (1976), Asmussen and Feldman (1977), Guckenheimer (1977), Milnor and Thurston (1977), Misiurewicz and Szlenk (1977), Štefan (1977), Feigenbaum (1978) and (1979).

3. WELL-BEHAVED PIECEWISE MONOTONE FUNCTIONS

For $f \in \mathcal{M}([a,b])$ let $A(f)$ denote the set of points in $[a,b]$ which are attracted to some periodic orbit of f, i.e.

$$A(f) = \{ y \in [a,b] : y \in A([x],f) \text{ for some } [x] \in \mathcal{P}(f) \} .$$

It is easy to see that we also have

$$A(f) = \{ x \in [a,b] : \lim_{m\to\infty} f^{nm}(x) \text{ exists for some } n \geq 1 \} ,$$

and thus in some sense $A(f)$ consists of those points y in $[a,b]$ for which the orbit $\{f^n(y)\}_{n\geq 0}$ has a particularly simple behaviour.
The starting point for what we will study in this section is the question: For which functions $f \in \mathcal{M}([a,b])$ does $A(f)$ contain a dense open subset of $[a,b]$? We can phrase this question more informally as: For which functions $f \in \mathcal{M}([a,b])$ does a typical point in $[a,b]$ get attracted to a periodic orbit ?

Remark: We have considered here that a subset B of $[a,b]$ contains a "typical" point of $[a,b]$ if B contains a dense open subset of $[a,b]$ (i.e. if the interior of B is dense in $[a,b]$). The main reason for this definition of "typical" is that it is the easiest to work with. Later in these notes (in Section 9) we will consider the more difficult question: For which functions $f \in \mathcal{S}$ does $A(f)$ have Lebesgue measure one?

The basic result of this section is Theorem 3.1. This will give us at least a partial answer to the above question, and it will also provide some additional information which will be especially useful in dealing with functions in $\mathcal{S}$. In order to state Theorem 3.1 we need a definition.

Let $f \in \mathcal{M}([a,b])$ and $J \subset [a,b]$ be a closed, non-trivial interval; J is called a *sink* of f if for some $n \geq 1$ we have f^n is monotone on J and $f^n(J) \subset J$. Note that if J is a sink of f then f^m is monotone on J for all $m \geq 0$; also we have $J \subset A(f)$. (Let $n \geq 1$ be such that f^n is monotone on J and $f^n(J) \subset J$; then for each $x \in J$ the sequence $\{f^{nm}(x)\}_{m \geq 0}$ is monotone and thus $\lim_{m \to \infty} f^{nm}(x)$ exists.) Let

$$S(f) = \{ x \in [a,b] : f^m(x) \in \operatorname{int}(J) \text{ for some sink } J \text{ and } m \geq 0 \},$$

(where $\operatorname{int}(B)$ denotes the interior of the set B). $S(f)$ is clearly an open subset of $[a,b]$ and $S(f) \subset A(f)$. (If $x \in S(f)$ then $f^m(x) \in A(f)$ for some $m \geq 0$; but if $f^m(x) \in A(f)$ then we also have $x \in A(f)$.)

Theorem 3.1 Let $f \in \mathcal{M}([a,b])$ satisfy

(3.1) f has a continuous second derivative in $(a,b)-T(f)$, and $f'(x) \neq 0$ for all $x \in (a,b)-T(f)$.

Suppose that for each $w \in T(f)$ there exist $\delta > 0$, $m \geq 0$ and a sink J such that $f^m([w-\delta,w+\delta]) \subset J$. Then $S(f)$ is a dense subset of $[a,b]$ (and so in particular $A(f)$ contains a dense open subset of $[a,b]$).

Proof Later. ◻◻

Before we investigate when the hypotheses of Theorem 3.1 are satisfied let us examine some of the consequences of $S(f)$ being a dense subset of $[a,b]$. For $f \in \mathcal{M}([a,b])$ let

$$A_s(f) = \{ y \in [a,b] : y \in \operatorname{int}(A([x],f)) \text{ for some } [x] \in \mathcal{P}_s(f) \}$$

and $L(f) = \{ y \in [a,b] : y \in \mathrm{int}(Per(n,f))$ for some $n \geq 1 \}$; also let $A_\ell(f) = \{ y \in [a,b] : f^m(y) \in L(f)$ for some $m \geq 0 \}$.

$A_s(f)$ and $A_\ell(f)$ are clearly both open and $A_s(f) \cup A_\ell(f) \subset A(f)$. Note that $L(f) \neq \phi$ (and hence $A_\ell(f) \neq \phi$) if and only if there exists $n \geq 1$ and a non-trivial interval $J \subset [a,b]$ with $f^n(x) = x$ for all $x \in J$.

Proposition 3.1 Let $f \in \mathcal{M}([a,b])$; then the interior of the set $S(f)-(A_s(f) \cup A_\ell(f))$ is empty. In particular, if $S(f)$ is a dense subset of $[a,b]$ then so is $A_s(f) \cup A_\ell(f)$.

Proof Let J be a sink of f and $n \geq 1$ be such that f^n is monotone on J and $f^n(J) \subset J$. It is easy to see that

$$J-(A_s(f) \cup A_\ell(f)) \subset \partial(\{ x \in J : f^n(x) = x \})$$

(where ∂B denotes the boundary of the set B), and therefore $J-(A_s(f) \cup A_\ell(f))$ has no interior. Now suppose that the interior of $S(f)-(A_s(f) \cup A_\ell(f))$ is not empty; there thus exists a non-trivial interval $I \subset S(f)-(A_s(f) \cup A_\ell(f))$. Take $x \in I$; then we can find $m \geq 0$ and a sink J such that $f^m(x) \in \mathrm{int}(J)$. Let $K = f^m(I) \cap J$, so K is a non-trivial interval and we have $K \subset J-(A_s(f) \cup A_\ell(f))$ (since if $y \notin A_s(f) \cup A_\ell(f)$ then $f^m(y) \notin A_s(f) \cup A_\ell(f)$). This is not possible and hence $S(f)-(A_s(f) \cup A_\ell(f))$ has no interior. Finally, suppose $S(f)$ is dense in $[a,b]$, and let F be any non-empty open subset of $[a,b]$. $S(f) \cap F$ is then a non-empty open subset of $[a,b]$ and so we cannot have $S(f) \cap F \subset S(f)-(A_s(f) \cup A_\ell(f))$. Thus $F \cap (A_s(f) \cup A_\ell(f))$ is non-empty, and this shows that $A_s(f) \cup A_\ell(f)$ is also a dense subset of $[a,b]$. ∎∎

Proposition 3.1 tells us that if the hypotheses of Theorem 3.1 are satisfied and if $L(f) = \phi$ then $A_s(f)$ is a dense subset of $[a,b]$; thus in this case a "typical" point in $[a,b]$ is attracted to a stable or a one-sided stable periodic orbit. For most functions f which occur in practice we do have $L(f) = \phi$, and this will often follow from the next result.

Proposition 3.2 Let $f \in M([a,b])$ have at least one turning point. Suppose f has a continuous derivative in $[a,b]$ and that f is real analytic in $(a,b)-T(f)$. Then $L(f) = \phi$.

Proof Suppose $L(f) \neq \phi$; then there exists $n \geq 1$ and a non-trivial interval $J \subset [a,b]$ such that $f^n(x) = x$ for all $x \in J$. Let $[u,v]$ be the largest interval containing J on which f^n is increasing; thus f^n is real analytic on (u,v) and hence $f^n(x) = x$ for all $x \in [u,v]$. In particular we have $(f^n)'(x) = 1$ for all $x \in (u,v)$ and thus also for all $x \in [u,v]$. Now we cannot have $[u,v] = [a,b]$ (since $T(f) \neq \phi$) and so if we let $w = \begin{cases} u & \text{if } u > a , \\ v & \text{otherwise,} \end{cases}$ then $w \in T(f^n)$. This gives us a contradiction because $(f^n)'(w) = 0$ for all $w \in T(f^n)$, and therefore we must have $L(f) = \phi$. ∎∎

We now apply Theorem 3.1 to functions in $\mathcal{S}$, and to do this it will be convenient to split things up into three cases. The first case is trivial: this is when we have $f \in \mathcal{S}$ with $f(\varphi) \leq \varphi$. Here we have $S(f) \subset [0,\varphi) \cup (\varphi,1]$ (noting that $[0,\varphi]$ is a sink) and in fact $A(f) = [0,1]$. (For each $x \in [0,1]$ the sequence $\{f^m(x)\}_{m\geq 1}$ is monotone

(3.4) $\varphi \in A([x],f)$ for some stable $[x] \in P_s(f)$,

(3.5) $\varphi \in A([x],f)$ for some one-sided stable $[x] \in P_s(f)$ but $f^k(\varphi) \neq x$ for all $k \geq 0$.

Then $S(f)$ is a dense subset of $[0,1]$.

Proof Again by Theorem 3.1 it is enough to show that there exist $\delta > 0$, $m \geq 0$ and a sink J such that $f^m([\varphi-\delta,\varphi+\delta]) \subset J$. First suppose (3.3) holds and let $[x]$ be the element of $P_s(f)$ which is neither trapped nor a fixed point of f in $[0,\gamma]$; thus $[x]$ is either critical or free and non-exceptional. If $[x]$ is critical then it is stable and of course $\varphi \in A([x],f)$; hence (3.4) holds and so we can deal with this case below. Assume then that $[x]$ is free and non-exceptional and that x is of period n ; for $0 \leq k < n$ let $[u_k,v_k]$ be the largest interval containing $f^k(x)$ on which f^{2n} is increasing. The proof of Proposition 2.6 shows that for some $0 \leq j < n$ either $u_j \neq 0$ and $f^{2n}(z) > z$ for all $z \in (u_j,f^j(x))$ or $v_j \neq 1$ and $f^{2n}(z) < z$ for all $z \in (f^j(x),v_j)$. Without loss of generality suppose the former holds and put $J = [u_j,f^j(x)]$; J is thus a sink. But $u_j \in T(f^{2n})$ and so by Proposition 2.1 there exists $0 \leq k < 2n$ with $f^k(u_j) = \varphi$. It is now easy to check that $f^{2n-k}([\varphi-\delta,\varphi+\delta]) \subset J$ for some $\delta > 0$. Next we consider (3.4) and (3.5). Let $[x] \in P_s(f)$ and suppose that x has period n ; let $[u,v]$ be the largest interval containing x such that f^{2n} is increasing on $[u,v]$, $f^{2n}(z) < z$ for all $z \in (x,v)$ and $f^{2n}(z) > z$ for all $z \in (u,x)$. It is easily seen that $[u,v]$ is non-trivial, and therefore $[u,v]$ is a sink. Furthermore, if $[x]$ is stable and non-critical then $x \in (u,v)$; thus if $z \in A([x],f)$ then

$f^m(z) \in (u,v)$ for some $m \geq 0$. Similarly, if $[x]$ is one-sided stable and $z \in A([x],f)$ is such that $f^k(z) \neq x$ for all $k \geq 0$ then we again have $f^m(z) \in (u,v)$ for some $m \geq 0$ (although this time x is an end-point of $[u,v]$). But if $f^m(z) \in (u,v)$ then for some $\delta > 0$ we have $f^m([z-\delta,z+\delta]) \subset [u,v]$. Finally, if $[x]$ is critical and $z \in A([x],f)$ then $f^m(z) \in [u,v]$ for some $m \geq 0$, and in this case we have $f^{m+1}([z-\delta,z+\delta]) \subset [u,v]$ for some $\delta > 0$. ▫▫

In practically all cases where (3.3) holds one of (3.4) and (3.5) will also hold: Let $[x]$ be an element of $P_s(f)$ which is neither trapped nor a fixed point of f in $[0,\gamma]$. Then, as noted in the proof of Theorem 3.3, $[x]$ is either critical or free and non-exceptional, and if $[x]$ is critical then (3.4) holds. If $[x]$ is free and non-exceptional then by Proposition 2.6 we know that $\varphi \in A([x],f)$ unless φ is an unstable periodic point. Thus if φ is not an unstable periodic point then either (3.4) or (3.5) will hold except in the very exceptional case where $[x]$ is one-sided stable and $f^k(\varphi) = x$ for some $k \geq 0$. The main reason for including (3.3) as a possible hypothesis in Theorem 3.3 is for the case when we know that φ is not an unstable periodic point. In this case (3.3) implies that φ is either regular or a stable periodic point. Thus if in Theorem 3.3 we have the additional hypothesis that any critical orbit of f is stable (which, for example, will always be true when f has a continuous derivative in $[0,1]$) then we can replace (3.3) by

(3.6) φ is regular.

Note that (3.4), (3.5) and (3.6) are all statements about the orbit $\{f^n(\varphi)\}_{n \geq 0}$. Theorem 3.3 says roughly that if this orbit is well-behaved then the orbit $\{f^n(x)\}_{n\geq 0}$ of a "typical" point x in $[0,1]$ is also well-behaved.

Let us now specialize even further and see what Theorems 3.2 and 3.3 give us for a particular subclass of functions in $\mathcal{S}$.

Theorem 3.4 Let $f \in \mathcal{S}$ with $f(x) > x$ for all $x \in (0,\varphi]$. Suppose f has a continuous derivative in $[0,1]$ and $f'(x) \neq 0$ for all $x \in (0,\varphi)\cup(\varphi,1)$; suppose also that f is real analytic in $(0,\varphi)\cup(\varphi,1)$ and that all trapped orbits of f are unstable. Then $|P_s(f)| \leq 1$ and $|P_s(f)| = 1$ if and only if φ is either regular or periodic. Moreover:

(1) If $|P_s(f)| = 1$ and $[x] \in P_s(f)$ then $A([x],f)$ contains a dense open subset of $[0,1]$ (and so in particular $A(f)$ contains a dense open subset of $[0,1]$).

(2) If $|P_s(f)| = 0$ then $A(f)$ is countable.

Proof Theorem 2.1(2) implies that $|P_s(f)| \leq 1$ (since $P_s(f)$ contains no trapped orbits); also Proposition 2.7 and the discussion following it show that $|P_s(f)| = 1$ if and only if φ is either regular or periodic. (Note that any critical orbit of f is stable because we are assuming f has a continuous derivative in $[0,1]$.) Suppose now that $|P_s(f)| = 1$ and let $[x] \in P_s(f)$. Then either (3.4) or (3.6) holds and so by Theorem 3.3 $S(f)$ is dense; thus by Proposition 3.1 $A_s(f) \cup A_\ell(f)$ is dense in $[0,1]$. But Proposition 3.2 gives us that $A_\ell(f) = \phi$ and of course $A_s(f) = \mathrm{int}(A([x],f))$; therefore $A([x],f)$ contains a dense open subset of $[0,1]$. Finally, suppose $|P_s(f)| = 0$; then $P(f) = P_u(f)$. Now as we have already remarked in Section 2, if $[x] \in P_u(f)$ then $A([x],f) = \{ y \in [0,1] : f^k(y) = x$ for some $k \geq 0 \}$, and thus $A([x],f)$ is countable. It is therefore enough to show that

$\mathcal{P}(f)$ is countable. Let $n \geq 1$ and $[u,v]$ be a maximal interval on which f^n is increasing; thus f^n is real analytic in (u,v). Exactly as in the proof of Proposition 3.2 we cannot have $f^n(x) = x$ for all $x \in (u,v)$ and hence $\{ z \in [u+\delta, v-\delta] : f^n(z) = z \}$ is finite for each $\delta > 0$. Therefore $\mathrm{Fix}(f^n) \cap [u,v]$ is countable and thus $\mathrm{Fix}(f^n)$ is also countable (since $\mathrm{Fix}(f^n)$ contains at most one element in each interval on which f^n is decreasing). This clearly implies that $\mathcal{P}(f)$ is countable. □□

Note that if $f \in \mathcal{S}$ satisfies the hypotheses of the above theorem then $A(f)$ is either "very large" (contains a dense open subset of $[0,1]$) or "very small" (is countable).

Theorem 3.4 corresponds to Theorem 3.3; we next give an analogous result which corresponds to Theorem 3.2.

Theorem 3.5 Let $f \in \mathcal{S}$ with $f(\varphi) > \varphi$ and $f^2(\varphi) < \gamma$. Suppose (3.2) holds, that $L(f) = \phi$ and that all trapped orbits of f are unstable. Then $\{ x \in [0,1] : f^n(x) \in [0,\gamma) \text{ for some } n \geq 0 \}$ is a dense open subset of $[0,1]$.

Note: If $f^n(x) \in [0,\gamma)$ then $f^m(x) \in [0,\gamma)$ for all $m \geq n$. Thus if the hypotheses of this theorem are satisfied then the orbit of a "typical" point in $[0,1]$ is eventually trapped in the interval $[0,\gamma)$.

Proof By Theorem 3.2 we know that $S(f)$ is dense in $[0,1]$ and so it is enough to show that

$$S(f) = \{ x \in [0,1] : f^n(x) \in [0,\gamma) \text{ for some } n \geq 0 \}.$$

Let J be a sink of f; then, since $L(f) = \phi$, it is easy to see that $x \in J$ for some $[x] \in \mathcal{P}_s(f)$. But Theorem 2.2(2) and the assumption that all trapped orbits are unstable imply that $P_s(f)$ consists of at most a fixed point of f in $[0,\gamma]$; thus $J \cap [0,\gamma] \neq \phi$. However, for each $\delta > 0$ there exists $y \in (\gamma,\gamma+\delta)$ and $k \geq 0$ with $f^k(y) = \varphi$, and on the other hand f^m is monotone on J for all $m \geq 0$; therefore $J \subset [0,\gamma]$. This, together with the fact that $[0,\gamma]$ is itself a sink, gives us what we want. □□

Remark: If $f^n(x) \in [0,\gamma)$ for some $n \geq 0$ then in fact $\lim_{m\to\infty} f^m(x)$ exists (and is a fixed point of f in $[0,\gamma]$). Thus when the hypotheses of Theorem 3.5 are satisfied we have that

$$\{ x \in [0,1] : \lim_{m\to\infty} f^m(x) \text{ exists} \}$$

contains a dense open subset of $[0,1]$.

The main problem in applying Theorems 3.4 and 3.5 comes in checking the assumption that all the trapped orbits of f are unstable. As we have already mentioned, we will deal with this problem in the next section.

We now return to the more general situation and give what is really the generalization of Theorem 3.3 to functions in $M([a,b])$.

Theorem 3.6 Let $f \in M([a,b])$ satisfy (3.1) and suppose that any critical orbit of f is stable; suppose also that for each $w \in T(f)$ one of the following three conditions is satisfied:

(3.7) $w \in A([x],f)$ for some stable $[x] \in P_s(f)$,

(3.8) $w \in A([x],f)$ for some one-sided stable $[x] \in P_s(f)$ but $f^k(w) \neq x$ for all $k \geq 0$,

(3.9) w is regular.

Then $S(f)$ is a dense subset of $[a,b]$.

Proof This follows exactly as in the proof of Theorem 3.3. ∎∎

Before starting the proof of Theorem 3.1 let us give another consequence of $S(f)$ being a dense open subset of $[a,b]$. Let $f \in C([a,b])$ and $x \in [a,b]$; we say that f is *orbit continuous* at x if for each $\varepsilon > 0$ there exists $\delta > 0$ such that $|f^n(y)-f^n(x)| < \varepsilon$ for all $n \geq 0$ and for all $y \in [a,b]$ with $|y-x| < \delta$. Let $O(f)$ denote the set of points in $[a,b]$ at which f is orbit continuous.

Proposition 3.3 If $f \in M([a,b])$ then $S(f) \cap \mathrm{int}(O(f))$ is dense in $S(f)$. In particular, if $S(f)$ is dense in $[a,b]$ then $O(f)$ contains a dense open subset of $[a,b]$.

Proof Let J be a sink of f and $n \geq 1$ be such that f^n is monotone on J and $f^n(J) \subset J$. Then it is not difficult to check that $J-O(f) \subset \partial(\{ x \in J : f^n(x) = x \})$, and therefore $J \cap \mathrm{int}(O(f))$ is dense in J . The rest of the proof is now the same as the proof of Proposition 3.1. ∎∎

If $f \in M([a,b])$ is such that $S(f)$ is dense in $[a,b]$ then by Proposition 3.3 (and the fact that $S(f) \subset A(f)$) we have $A(f) \cap O(f)$ contains a dense open subset of $[a,b]$. Thus if $S(f)$ is dense then the iterates of f are well-behaved in the sense that

(i) for each point x in a "large" subset of $[a,b]$ the orbit $\{f^n(x)\}_{n\geq 0}$ has a very simple behaviour,

(ii) on a "large" subset of $[a,b]$ the orbits $\{f^n(x)\}_{n\geq 0}$ vary continuously as a function of x .

We now start the proof of Theorem 3.1. Fix $f \in \mathcal{M}([a,b])$ and let

$$H = \{ x \in [a,b] : \text{there exists } \delta > 0 \text{ such that } f^m(y) \notin T(f) \text{ for all } m \geq 0 \text{ and for all } y \in [a,b] \text{ with } |y-x| < \delta \} .$$

Thus H is open, $f(H) \subset H$, and $[a,b]-H = \overline{M}$, where

$$M = \{ x \in [a,b] : f^n(x) \in T(f) \text{ for some } n \geq 0 \} ,$$

(and where $\overline{B}$ denotes the closure of the set B). Theorem 3.1 will follow from a careful analysis of the set H .

Lemma 3.1 If J is a sink of f then $\mathrm{int}(J) \subset H$.

Proof This is clear, because f^m is monotone on J for all $m \geq 0$. ⊞

Lemma 3.1 shows that if f satisfies the hypotheses of Theorem 3.1 then $H \neq \phi$. Thus from now on let us assume that H is non-empty. We can write $H = \bigcup_{t \in V} H_t$ as a countable disjoint union of non-empty open intervals. If we let $\mathcal{J}$ denote the set of open intervals J such that $f^m(x) \notin T(f)$ for all $x \in J$, $m \geq 0$ then it is not hard to see that the H_t are exactly the maximal elements in $\mathcal{J}$. They are also the maximal open intervals J with the property that f^m is monotone on J for each $m \geq 0$.

Let $t \in V$, $n \geq 0$; then $f^n(H_t)$ is a connected subset of H and and so there exists (a unique) $t(n) \in V$ with $f^n(H_t) \subset H_{t(n)}$. We have $H_{(t(m))(n)} \supset f^n(H_{t(m)}) \supset f^n(f^m(H_t)) = f^{n+m}(H_t)$, and thus

(3.10) $(t(m))(n) = t(n+m)$ for all $t \in V$, $n, m \geq 0$.

We say that $t \in V$ is *periodic* if $f^n(H_t) \subset H_t$ for some $n \geq 1$; the smallest such $n \geq 1$ is called the *period* of t . Clearly $t \in V$ is periodic if and only if $t(n) = t$ for some $n \geq 1$, and the smallest $n \geq 1$ with this property is then the period of t . If $t \in V$ is periodic then (3.10) shows that $t(m)$ is also periodic for each $m \geq 0$, and that t and $t(m)$ have the same period. We call $t \in V$ *eventually periodic* if $t(n)$ is periodic for some $n \geq 0$ (and so in particular t is eventually periodic if it is periodic); let V_0 denote the set of eventually periodic elements of V . It is not hard to see that $t \in V-V_0$ if and only if $t(n) \neq t(m)$ for all $0 \leq n < m$.

Lemma 3.2 $S(f) = \bigcup_{t \in V_0} H_t$.

Proof If $t \in V$ is periodic with period n then $f^n(H_t) \subset H_t$ and f^n is monotone on H_t ; thus $\overline{H}_t$ is a sink and hence $\bigcup_{t \in V_0} H_t \subset S(f)$.

Conversely, let J be a sink; by Lemma 3.1 we have $int(J) \subset H$ and so let $t \in V$ be such that $int(J) \subset H_t$. Now for some $n \geq 1$ we have $f^n(int(J)) \subset int(J)$ and thus also $f^n(H_t) \cap H_t \neq \phi$. But this implies that $f^n(H_t) \subset H_t$ and therefore t is periodic. From this it easily follows that $S(f) \subset \bigcup_{t \in V_0} H_t$. □□

Lemma 3.2 reduces the proof of Theorem 3.1 to showing that $V_0 = V$ and H is dense in $[a,b]$.

Lemma 3.3 H is a dense subset of $[a,b]$ if and only if $w \notin int(\overline{M})$ for each $w \in T(f)$.

Proof If $w \in \text{int}(\overline{M})$ for some $w \in T(f)$ then in particular $\text{int}(\overline{M}) \neq \phi$ and so H is not dense. Conversely, suppose that H is not dense; then there exists a non-empty open interval $J \subset (a,b)$ with $J \cap H = \phi$. We thus have $J \subset \overline{M}$ and hence we can find $y \in J \cap M$. Let $m \geq 0$ be the smallest integer with $f^m(y) \in T(f)$ and put $w = f^m(y)$; we will show that $w \in \text{int}(\overline{M})$. Now f^m is strictly monotone in a neighbourhood of y and so $w \in \text{int}(f^m(J))$; thus it is enough to show that $f^m(J) \subset \overline{M}$. Let $z \in J$; then since $J \subset \overline{M}$ we can find $z_n \in M$, $n \geq 1$, with $z_j \neq z_k$ for all $1 \leq j < k$ and such that $z = \lim_{n\to\infty} z_n$. However, it is clear that $\{ y \in [a,b] : f^k(y) \in T(f) \text{ for some } k < m \}$ is finite and also that $f^m(y) \in M$ provided $f^k(y) \in T(f)$ for some $k \geq m$. Thus all but finitely many members of the sequence $\{f^m(z_n)\}_{n\geq 1}$ are in M and therefore $f^m(z) \in \overline{M}$. This gives us $f^m(J) \subset \overline{M}$, which is what we wanted. ▫▫

Lemma 3.3 shows in particular that if each $w \in T(f)$ is isolated in M (i.e. if for each $w \in T(f)$ there exists $\delta > 0$ so that $(w-\delta,w+\delta)-H = \{w\}$) then H is dense in $[a,b]$.

Lemma 3.4 Let $w \in T(f)$ and suppose there exist $\delta > 0$, $m \geq 0$ and a sink J such that $f^m([w-\delta,w+\delta]) \subset J$. Then w is isolated in M.

Proof Since $\{ y \in [a,b] : f^k(y) \in T(f) \text{ for some } k < m \}$ is finite we can find $\varepsilon > 0$ such that f^k is monotone on both $(w-\varepsilon,w)$ and $(w,w+\varepsilon)$ for all $k < m$ and such that $f^m((w-\varepsilon,w)\cup(w,w+\varepsilon)) \subset \text{int}(J)$. We then have $(w-\varepsilon,w)\cup(w,w+\varepsilon) \subset H$ and thus w is isolated in M. ▫▫

The next result provides the final step in the proof of Theorem 3.1.

Proposition 3.4 If each $w \in T(f)$ is an isolated point of M and (3.1) holds then $V_0 = V$.

Proof We proceed via a couple of lemmas. Let $a < d_1 < \cdots < d_N < b$ be the turning points of f and for $\delta > 0$ let

$$I(\delta) = [a+\delta, d_1-\delta] \cup [d_1+\delta, d_2-\delta] \cup \cdots \cup [d_{N-1}+\delta, d_N-\delta] \cup [d_N+\delta, b-\delta] .$$

Lemma 3.5 If each $w \in T(f)$ is an isolated point of M and $V_0 \neq V$ then there exists $t \in V-V_0$ and $\delta > 0$ such that $H_{t(n)} \subset I(\delta)$ for all $n \geq 0$.

Proof Let $w \in T(f)$; then since w is isolated in M there exist $q(w), r(w) \in V$ such that w is the left-hand (resp. right-hand) end-point of $\overline{H}_{q(w)}$ (resp. $\overline{H}_{r(w)}$). We can thus find $\varepsilon > 0$ with $(w-\varepsilon, w+\varepsilon) \subset \overline{H}_{q(w)} \cup \overline{H}_{r(w)}$ for all $w \in T(f)$ and then also $\overline{\varepsilon} > 0$ such that $\overline{\varepsilon} < f(x) < 1-\overline{\varepsilon}$ for all $x \in I(\varepsilon)$. Let $s \in V-V_0$; then since $s(n) \neq s(m)$ for all $0 \leq n < m$ we can find $j \geq 0$ such that $s(n) \notin \bigcup_{w \in T(f)} \{q(w), r(w)\}$ for all $n \geq j$. Now if $0 < \overline{\delta} \leq \min\{\varepsilon, \overline{\varepsilon}\}$ and $n > j$ then $H_{s(n)} \not\subset I(\overline{\delta})$ can only hold if $H_{s(n-1)} \subset [a, d_1] \cup [d_N, b]$ and $H_{s(n)} \cap ([a, a+\overline{\delta}] \cup [b-\overline{\delta}, b]) \neq \phi$. From this it is not hard to see that there exists $0 < \delta \leq \min\{\varepsilon, \overline{\varepsilon}\}$ and $k \geq j$ such that $H_{s(n)} \cap ([a, a+\delta] \cup [b-\delta, b]) = \phi$ for all $n \geq k$. Therefore if we let $t = s(k)$ then $t \in V-V_0$ and $H_{t(n)} \subset I(\delta)$ for all $n \geq 0$. ⊞⊞

Let us now suppose that (3.1) holds; then for $\delta > 0$ we can let

$B(\delta) = \sup\{\ |f''(x)/f'(x)|\ :\ x \in I(\delta)\ \}$, and $B(\delta)$ is finite.

Lemma 3.6 Let $a < c < d < b$, $\delta > 0$, $\varepsilon > 0$ and $n \geq 1$; suppose that $f^k([c,d+\varepsilon]) \subset I(\delta)$ for all $k = 0,1,\ldots,n-1$. Then

$$(3.11) \qquad \frac{|f^n([d,d+\varepsilon])|}{|f^n([c,d])|} \leq \frac{\varepsilon}{d-c} \exp\left\{B(\delta) \sum_{k=0}^{n-1} |f^k([c,d+\varepsilon])|\right\} .$$

Proof If x and y are in the same component of $I(\delta)$ then by the mean value theorem we have $|\log|f'(x)|-\log|f'(y)|| \leq B(\delta)|x-y|$, and thus $|f'(x)|/|f'(y)| \leq \exp(B(\delta)|x-y|)$. Let $x, y \in [c,d+\varepsilon]$; for $k = 0,1,\ldots,n-1$ we have $f^k([c,d+\varepsilon]) \subset I(\delta)$ and so $f^k(x)$ and $f^k(y)$ are in the same component of $I(\delta)$. Hence

$$\frac{|f'(f^k(x))|}{|f'(f^k(y))|} \leq \exp(B(\delta)|f^k(x)-f^k(y)|) \leq \exp(B(\delta)|f^k([c,d+\varepsilon])|) ,$$

and therefore

$$\frac{|(f^n)'(x)|}{|(f^n)'(y)|} = \prod_{k=0}^{n-1} \frac{|f'(f^k(x))|}{|f'(f^k(y))|} \leq \exp\left\{B(\delta) \sum_{k=0}^{n-1} |f^k([c,d+\varepsilon])|\right\} .$$

This immediately gives us (3.11). ▫▫

Now we suppose that (3.1) holds, that each $w \in T(f)$ is an isolated point of M , but that $V_0 \neq V$. Let $t \in V - V_0$ and $\delta > 0$ be given by Lemma 3.5, and let $H_t = (c,d)$.

Lemma 3.7 There exists $\varepsilon > 0$ such that $f^n((c,d+\varepsilon)) \subset I(\delta/2)$ for all $n \geq 0$.

Proof Choose $\varepsilon > 0$ so that $\varepsilon < \min\{\delta/2, d-c\}$ and $\varepsilon < (d-c)[\min\left\{1, \frac{\delta}{2(b-a)}\right\}]\exp(-2(b-a)B(\delta/2))$. We will show by induction that $f^k((c,d+\varepsilon)) \subset I(\delta/2)$ and $|f^k([c,d+\varepsilon])| \leq 2|f^k([c,d])|$ for all $k \geq 0$. This is clearly true when $k = 0$ since $\varepsilon < \min\{\delta/2, d-c\}$; so suppose it holds for $k = 0,1,\ldots,n-1$. Note that the intervals $\{f^m((c,d))\}_{m\geq 0}$ are disjoint and $f^m((c,d)) \subset I(\delta)$ for each $m \geq 0$ (since $f^m((c,d)) = f^m(H_t) \subset H_{t(m)}$ and the $H_{t(m)}$ are disjoint subsets of $I(\delta)$). Thus

$$\sum_{k=0}^{n-1} |f^k([c,d+\varepsilon])| \leq 2 \sum_{k=0}^{n-1} |f^k([c,d])| \leq 2(b-a) ,$$

and so by Lemma 3.6

$$|f^n([d,d+\varepsilon])| \leq \frac{\varepsilon}{d-c} \exp(2(b-a)B(\delta/2))|f^n([c,d])| ;$$

therefore $|f^n([d,d+\varepsilon])| < \min\left\{1, \frac{\delta}{2(b-a)}\right\}|f^n([c,d])|$. This implies that $|f^n([c,d+\varepsilon])| \leq 2|f^n([c,d])|$, and it also gives us $f^n((c,d+\varepsilon)) \subset I(\delta/2)$ because $|f^n([d,d+\varepsilon])| < \delta/2$ and of course $f^n(d) \in f^n([c,d]) \subset I(\delta)$. ◻◻

However, the conclusion of Lemma 3.7 implies that if $\varepsilon > 0$ is small enough then $(c,d+\varepsilon) \subset H$, and this contradicts the fact that $H_t = (c,d)$ is a maximal connected component of H . Therefore we must have $V_o = V$, and this completes the proof of Proposition 3.4. ◻◻

Putting together Lemmas 3.2, 3.3 and 3.4 and Proposition 3.4 also completes the proof of Theorem 3.1. ☐☐

Remark: For any function $f \in \mathcal{M}([a,b])$ we have $\bigcup_{t \in V} H_t \subset O(f)$. (This follows because if $t \in V - V_0$ then the intervals $\{H_{t(n)}\}_{n \geq 0}$ are disjoint, and so $\lim_{n\to\infty} |H_{t(n)}| = 0$.) Therefore if $f \in \mathcal{M}([a,b])$ is such that for each $w \in T(f)$ there exist $\delta > 0$, $m \geq 0$ and a sink J with $f^m([w-\delta, w+\delta]) \subset J$ then $O(f)$ contains a dense open subset of $[a,b]$, even when (3.1) does not hold.

Notes: Proposition 3.4 is taken from the proof of Proposition 2.8 in Guckenheimer (1979) and, as Guckenheimer points out, the estimates needed in this proposition are similar to ones to be found in Denjoy (1932).

4. PROPERTY R AND NEGATIVE SCHWARZIAN DERIVATIVES

Most of the results of Sections 2 and 3 can be best applied to functions for which all trapped orbits are unstable. We now consider a class of functions for which this is true. If $w, x, y, z \in \mathbb{R}$ are distinct then put $R(w,x,y,z) = \dfrac{(z-w)(y-x)}{(z-y)(x-w)}$; this is usually called the *cross-ratio* of w, x, y and z . Let $a, b \in \mathbb{R}$ with $a < b$ and let $f : [a,b] \to \mathbb{R}$ be continuous and strictly monotone. We say f has *property R* if

$$(4.1) \qquad R(f(x_1),f(x_2),f(x_3),f(x_4)) > R(x_1,x_2,x_3,x_4)$$

whenever $a < x_1 < x_2 < x_3 < x_4 < b$. We will also consider a slightly weaker condition and say that f has *property R'* if

$$(4.2) \qquad R(f(x_1),f(x_2),f(x_3),f(x_4)) \geq R(x_1,x_2,x_3,x_4)$$

whenever $a < x_1 < x_2 < x_3 < x_4 < b$. Note that if (4.2) holds whenever $a < x_1 < x_2 < x_3 < x_4 < b$ then by continuity it also holds whenever $a \leq x_1 < x_2 < x_3 < x_4 \leq b$. Let $f \in M([a,b])$ have turning points $d_1 < d_2 < \cdots < d_N$, put $d_0 = a$ and $d_{N+1} = b$; we say that f has *property R* (resp. *property R'*) if the restriction of f to each of its laps $[d_k,d_{k+1}]$, $k = 0,1,\ldots,N$, has property R (resp. property R'). We let $M_R([a,b])$ (resp. $M_{R'}([a,b])$) denote the functions in $M([a,b])$ which have property R (resp. property R'); S_R and $S_{R'}$ will denote the corresponding functions in S .

At first sight it does not look very easy to check whether a given function f has one of these properties or not. However, if f is smooth enough then we can make use of the following result.

Proposition 4.1 Let $f : [a,b] \longrightarrow \mathbb{R}$ be continuous and have a third derivative in (a,b) ; suppose $f'(x) \neq 0$ for all $x \in (a,b)$ (and thus in particular f is strictly monotone). Then f has property R (resp. property R˙) if and only if $(Sf)(x) < 0$ for all $x \in (a,b)$ (resp. $(Sf)(x) \leq 0$ for all $x \in (a,b)$), where

$$(Sf)(x) = \left[\frac{f''(x)}{f'(x)}\right]' - \frac{1}{2}\left[\frac{f''(x)}{f'(x)}\right]^2 \quad \left(= \frac{f'''(x)}{f'(x)} - \frac{3}{2}\left[\frac{f''(x)}{f'(x)}\right]^2 \right).$$

Sf is called the *Schwarzian derivative* of f .

Proof We leave this to the end of the section. ◻◻

Remark: Suppose f satisfies the hypotheses of Proposition 4.1; let $g(x) = |f'(x)|^{-1/2}$. An elementary calculation then shows that $g'' = -\frac{1}{2} g\, Sf$, and thus $(Sf)(x) < 0$ for all $x \in (a,b)$ (resp. $(Sf)(x) \leq 0$ for all $x \in (a,b)$) if and only if $|f'|^{-1/2}$ is strictly convex (resp. is convex).

From Proposition 4.1 we immediately deduce the following: Suppose $f \in M([a,b])$ has a third derivative in $(a,b)-T(f)$ and that $f'(x) \neq 0$ for all $x \in (a,b)-T(f)$. Then $f \in M_R([a,b])$ (resp. $f \in M_{R^*}([a,b])$) if and only if $(Sf)(x) < 0$ (resp. $(Sf)(x) \leq 0$) for all $x \in (a,b)-T(f)$. Examples of functions in S_R therefore include:

(i) $f(x) = \mu x(1-x)$ with $0 < \mu \leq 4$; in this case we have $(Sf)(x) = -\frac{3}{2}(x-\frac{1}{2})^{-2}$;

(ii) $f(x) = \sin(\mu x)$ with $\frac{\pi}{2} < \mu \leq \pi$: $(Sf)(x) = -\mu^2\left(1+\frac{3}{2}\tan^2(\mu x)\right)$;

(iii) $f(x) = \mu x \exp(1-\mu x)$ with $\mu > 1$; a simple calculation gives

$$(Sf)(x) = -\frac{\mu^2[(\mu x)^2-4(\mu x)+6]}{2(1-\mu x)^2}\ , \text{ and } y^2-4y+6 > 0 \text{ for all } y \in \mathbb{R}\ .$$

If $f \in \mathcal{S}$ is linear on each of the intervals $[0,\varphi]$ and $[\varphi,1]$ then f is in $\mathcal{S}_{R'}$ but not in $\mathcal{S}_R$ (since $(Sf)(x) = 0$ for all $x \in (0,\varphi)\cup(\varphi,1)$).

The main results of this section will depend on two important facts about functions having property R and property R′, namely:

(1) If $f \in \mathbb{M}_{R'}([a,b])$ then any trapped orbit of f is unstable.

(2) If $f \in \mathbb{M}_R([a,b])$ then $\mathrm{Per}(n,f)$ is finite for each $n \geq 1$, and so in particular $L(f) = \phi$.

(1) and (2) will allow us to improve many of the results from Sections 2 and 3. Before stating the first such "improved" result let us recall a couple of definitions from the end of Section 2. Let $f \in \mathbb{M}([a,b])$ have turning points $d_1 < d_2 < \cdots < d_N$; put $d_0 = a$ and $d_{N+1} = b$, and for $0 \leq k \leq N$ let $I_k = [d_k,d_{k+1}]$; also let
$\Lambda = \{\ \{\sigma_m\}_{m\geq 0} : \sigma_m \in \{0,1,\dots,N\}$ for each $m \geq 0\ \}$. We defined $\sigma = \{\sigma_m\}_{m\geq 0} \in \Lambda$ to be a code for $x \in [a,b]$ if $f^m(x) \in I_{\sigma_m}$ for all $m \geq 0$. Let $G(f) = \{\ x \in [a,b] : f^m(x) \notin T(f)$ for all $m \geq 0\ \}$; recall also that $w \in T(f)$ was defined to be regular if $f(w) \in G(f)$ and if one of the two possible codes for w is periodic. (*Note:* In the part of Section 2 where we first introduced these terms we had assumed that any critical orbit of f is stable. However, the definitions which we have

repeated here make sense without this assumption.)

Proposition 4.2 If $f \in M_{R'}([a,b])$ then any trapped orbit of f is unstable. Moreover, if $[x] \in P_s(f)$ is free and non-exceptional then there exists a regular $w \in T(f)$ and $\delta > 0$ such that $(w-\delta, w+\delta) \subset A([x],f)$.

Proof Later. ▫▫

Proposition 4.2 corresponds to Proposition 2.6, and it will perform a similar rôle to that played by Proposition 2.6 in Section 2.

Theorem 4.1 Suppose $f \in M_{R'}([a,b])$ has N turning points. Then

$|P_s(f)| \leq N+\Delta$, where $\Delta = \begin{cases} 1 & \text{if } N \text{ is odd,} \\ 2 & \text{if } N \text{ is even;} \end{cases}$ and in fact

$|P_s(f)| \leq N(f)+\Delta$, where

$N(f) = |\{ w \in T(f) : w \text{ is either regular or a stable periodic point } \}|$.

Proof This follows from Proposition 4.2 in exactly the same way as Theorem 2.1 followed from Proposition 2.6. ▫▫

Theorem 4.1 is a version of Theorem 2.1(1); it is also similar to Proposition 2.7 and (like Proposition 2.7) it can be used to give a procedure for finding the elements of $P_s(f)$:

Take $w \in T(f)$ and check first whether it is a stable periodic point; if it is then $[w] \in P_s(f)$. If w is not periodic then check whether it is regular; if it is then (by Proposition 2.7) it is attracted to an element of $P_s(f)$. Apply this procedure to all the points in $T(f)$.

The proof of Theorem 4.1 shows that there are at most two elements of

$P_s(f)$ which are not obtained in this way. These possibly remaining elements are either exceptional or contained in $\{a,b\}$, and they can be easily found.

We now specialize and see what happens for functions in S which have property R'. The next result corresponds to Theorem 2.1(2).

Theorem 4.2 Let $f \in S_{R'}$; then $P_s(f)$ contains at most one element $[x]$ which is not a fixed point of f in $[0,\varphi]$. Moreover, such an $[x]$ exists if and only if $f(\varphi) > \varphi$ and φ is either regular or a stable periodic point, and if such an $[x]$ exists then $(\varphi-\delta,\varphi+\delta) \subset A([x],f)$ for some $\delta > 0$.

Proof If $[x] \in P_s(f)$ is not a fixed point of f in $[0,\varphi]$ then $[x]$ is either critical or free and non-exceptional. In the first case φ is a stable periodic point and thus $(\varphi-\delta,\varphi+\delta) \subset A([\varphi],f) = A([x],f)$ for some $\delta > 0$; in the second case Proposition 4.2 implies that φ is regular and again $(\varphi-\delta,\varphi+\delta) \subset A([x],f)$ for some $\delta > 0$. In both cases we have $\varphi \in A([x],f)$, and hence at most one such $[x]$ can exist; also it is clear that if such an $[x]$ exists then $f(\varphi) > \varphi$. It therefore remains to show that if $f(\varphi) > \varphi$ and φ is either regular or a stable periodic point then $P_s(f)$ does contain an element which is not a fixed point of f in $[0,\varphi]$. If φ is a stable periodic point then this is clear, so assume φ is regular. The proof of Proposition 2.7 then gives us $[x] \in P_s(f)$ such that the periodic code for φ is also a code for x. But if $\{\sigma_m\}_{m\geq 0}$ is the periodic code for φ then $\sigma_1 = 1$ (since $f(\varphi) > \varphi$); thus $f(x) \in [\varphi,1]$ and hence x is not a fixed point of f in $[0,\varphi]$. ◻◻

As a corollary to Theorem 4.2 we have the following sufficient condition for $P_s(f)$ to be empty.

Proposition 4.3 Let $f \in S_{R^-}$ with $f(y) > y$ for all $y \in (0,\varphi]$ and suppose $\varphi \in A([z],f)$ for some $[z] \in P_u(f)$. Then $|P_s(f)| = 0$. In particular, if φ itself is an unstable periodic point then $|P_s(f)| = 0$.

Proof This follows immediately from Theorem 4.2. ◻◻

The simplest case of a function $f \in S_{R^-}$ satisfying the hypotheses of Proposition 4.3 is (as with Theorem 2.2(1)) when we have $f(y) > y$ for all $y \in (0,\varphi]$, $f(0) = f(1) = 0$ and $f(\varphi) = 1$. (See the first picture after Theorem 2.2.) Specific instances of this are given by the functions $f(x) = 4x(1-x)$ and $f(x) = \sin(\pi x)$. Other examples of functions $f \in S_{R^-}$ with $|P_s(f)| = 0$ occur when $f(y) > y$ for all $y \in (0,\varphi]$ and $f^n(\varphi) = \beta$ for some $n \geq 1$, where β is the unique fixed point of f in $(\varphi,1)$. (See the second picture after Theorem 2.2.) φ is then attracted to β , and β is an unstable fixed point. (β has a unique code consisting of all 1's ; thus if $[\beta] \in P_s(f)$ then by Theorem 4.2 this code would also be the periodic code for φ , and hence we would have $f^m(\varphi) \in [\varphi,1]$ for all $m \geq 0$. But this is not possible since if $n \geq 1$ is the smallest integer with $f^n(\varphi) = \beta$ then $f^{n-1}(\varphi) \in [0,\varphi)$.) It is easy to see that each of the families of functions $f_\mu(x) = \mu x(1-x)$, $0 < \mu \leq 4$, $f_\mu(x) = \sin(\mu x)$, $\frac{\pi}{2} < \mu \leq \pi$, and $f_\mu(x) = \mu x \exp(1-\mu x)$, $\mu > 1$, contains an infinite sequence of functions of this type.

We next turn our attention to the results in Section 3 and give new versions of some of these for functions having property R . The reason for now using property R (rather than just property R′) is that we want to have $L(f) = \phi$ and in general this is not true if we only have $f \in M_{R'}([a,b])$. However, the following proposition shows that it does hold for functions in $M_R([a,b])$.

Proposition 4.4 If $f \in M_R([a,b])$ then $Per(n,f)$ is finite for each $n \geq 1$; in fact if f has N turning points then for each $n \geq 1$ we have $|Fix(f^n)| \leq 3(N+1)^n$.

Proof Later. ◻◻

Theorem 4.3 Let $f \in S_R$ with $f(x) > x$ for all $x \in (0,\varphi]$ and suppose

(4.3) f has a continuous second derivative in $(0,\varphi)\cup(\varphi,1)$ and $f'(x) \neq 0$ for all $x \in (0,\varphi)\cup(\varphi,1)$.

Then $|P_s(f)| \leq 1$ and $|P_s(f)| = 1$ if and only if φ is either regular or a stable periodic point. Moreover:

(1) If $|P_s(f)| = 1$ and $[x] \in P_s(f)$ then $A([x],f)$ contains a dense open subset of $[0,1]$ (and so in particular $A(f)$ contains a dense open subset of $[0,1]$).

(2) If $|P_s(f)| = 0$ then $A(f)$ is countable.

Proof By Theorem 4.2 we have $|P_s(f)| \leq 1$ and also $|P_s(f)| = 1$ if and only if φ is either regular or a stable periodic point. Thus suppose that $|P_s(f)| = 1$ and let $[x] \in P_s(f)$. Then either (3.4) or (3.6) holds and therefore Theorem 3.4 and Proposition 3.1 imply $A_s(f) \cup A_\ell(f)$ is dense in $[0,1]$. But Proposition 4.4 gives us that $A_\ell(f) = \phi$ and of course $A_s(f) = int(A([x],f))$; hence $A([x],f)$ contains a dense open subset of $[0,1]$. The last part of the proof is the same as in Theorem 3.4 because Proposition 4.4 implies $P(f)$ is countable. ⊞

It should be clear that a slight modification of the above proof will produce the following version of Theorem 3.3: Suppose we have $f \in S_R$ with $f(\varphi) > \varphi$ and $f^2(\varphi) \geq \gamma$ (where $\gamma = \gamma(f)$ is defined as in Section 3), and that (4.3) holds. Then *(i)* $P_s(f)$ contains at most one element $[x]$ which is not a fixed point of f in $[0,\varphi]$; *(ii)* such an $[x]$ exists if and only if φ is either regular or a stable periodic point; and *(iii)* if $[x]$ exists then $A([x],f)$ contains a dense open subset of $[\gamma,\bar{\gamma}]$.

We next consider a result which corresponds to Theorem 3.5, and in order to state this we require the following fact.

Proposition 4.5 Let $f \in S_R$ with $f(\varphi) > \varphi$ and suppose $\gamma > 0$; let z be the smallest fixed point of f in $[0,\gamma]$. Then $f(y) > y$ for all $y \in [0,z)$ and $f(y) < y$ for all $y \in (z,\gamma)$.

Proof This will be a simple consequence of Proposition 4.6. ⊞

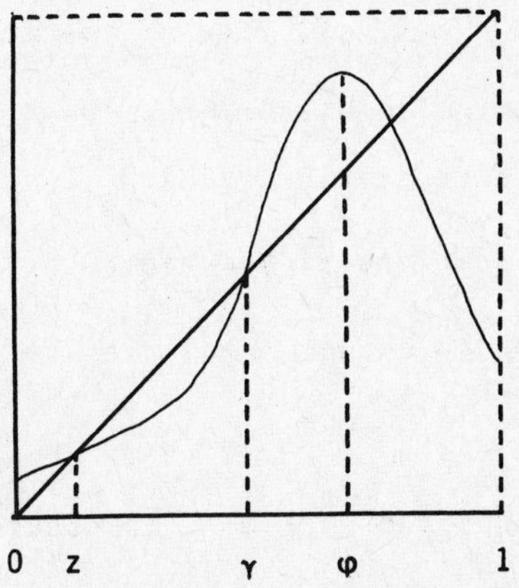

Remark: It is possible that z is either 0 or γ .

Theorem 4.4 Let $f \in S_R$ with $f(\varphi) > \varphi$ and $f^2(\varphi) < \gamma$; suppose that (4.3) holds. Then $P_s(f) = \{[z]\}$ (where z is again the smallest fixed point of f in $[0,\gamma]$) and $A([z],f)$ contains a dense open subset of $[0,1]$.

Proof φ is neither regular nor a stable periodic point and thus Theorem 4.2 implies that all the elements of $P_s(f)$ are fixed points of f in $[0,\varphi]$; this together with Proposition 4.5 gives us $P_s(f) = \{[z]\}$. The rest of the proof is now the same as in Theorem 4.3 because Theorem 3.2 tells us that $S(f)$ is dense in $[0,1]$. ◻◻

The final "improvement" of a result from Section 3 corresponds to Theorem 3.6; we thus return to the more general situation and again consider functions in $M([a,b])$.

Theorem 4.5 Let $f \in M_R([a,b])$ satisfy

(4.4) f has a continuous second derivative in $(a,b)-T(f)$ and $f'(x) \neq 0$ for all $x \in (a,b)-T(f)$.

Suppose also that for each $w \in T(f)$ one of the following three conditions is satisfied:

(4.5) $w \in A([x],f)$ for some stable $[x] \in P_s(f)$,

(4.6) $w \in A([x],f)$ for some one-sided stable $[x] \in P_s(f)$ but $f^k(w) \neq x$ for all $k \geq 0$,

(4.7) w is regular.

Then $A_s(f)$ is a dense subset of $[a,b]$ (and thus a "typical" point in [a,b] is attracted to a stable or a one-sided stable periodic orbit).

Proof This follows directly from Theorem 3.6 because Proposition 4.4 implies $A_\ell(f) = \phi$. ▫▫

Remark: If f has N turning points then by Theorem 4.1 we have $|P_s(f)| \leq N+\Delta$; thus $P_s(f) = \{[x_1],\dots,[x_M]\}$ for some $0 \leq M \leq N+\Delta$ and then $A_s(f) = \bigcup_{k=1}^{M} \mathrm{int}(A([x_k],f))$.

We now begin the analysis required for the proofs of Propositions 4.2, 4.4 and 4.5; this of course involves the study of functions having property R´ and property R.

Lemma 4.1 Let $f : [a,b] \to [c,d]$ and $g : [c,d] \to \mathbb{R}$ be continuous, strictly monotone functions. If f and g both have property R´ (resp. property R) then so does $g \circ f$.

Proof Suppose f and g have property R´ and let $a < x_1 < x_2 < x_3 < x_4 < b$; then either $c < f(x_1) < \cdots < f(x_4) < d$ or $c < f(x_4) < \cdots < f(x_1) < d$. In both cases we have

$$R((g\circ f)(x_1),(g\circ f)(x_2),(g\circ f)(x_3),(g\circ f)(x_4))$$

$$\geq R(f(x_1),f(x_2),f(x_3),f(x_4)) \geq R(x_1,x_2,x_3,x_4) ,$$

because $R(z_1,z_2,z_3,z_4) = R(z_4,z_3,z_2,z_1)$; this shows that $g \circ f$ has property R´. The other part follows in the same way. ▫▫

Lemma 4.2 $M_R([a,b])$ and $M_{R'}([a,b])$ are closed under composition, i.e. if f and g are both in $M_R([a,b])$ (resp. $M_{R'}([a,b])$) then so is

$g \circ f$.

Proof This is a straightforward consequence of Lemma 4.1. ▫▫

Lemma 4.2 shows in particular that if $f \in M_{R^-}([a,b])$ (resp. $f \in M_R([a,b])$) then also $f^n \in M_{R^-}([a,b])$ (resp. $f^n \in M_R([a,b])$) for all $n \geq 1$.

Proposition 4.6 Let $g : [c,d] \to \mathbb{R}$ be continuous, strictly increasing, and have property R^-; let $\ell, m \in \mathbb{R}$ and put $h(x) = \ell x + m$ (so h is linear). Suppose we have $c \leq y_1 < y_2 < y_3 \leq d$ with $g(y_1) \leq h(y_1)$, $g(y_2) = h(y_2)$ and $g(y_3) \geq h(y_3)$. Then $g(y) \leq h(y)$ for all $y \in (y_1,y_2)$ and $g(y) \geq h(y)$ for all $y \in (y_2,y_3)$. Moreover, if g has property R then $g(y) < h(y)$ for all $y \in (y_1,y_2)$ and $g(y) > h(y)$ for all $y \in (y_2,y_3)$.

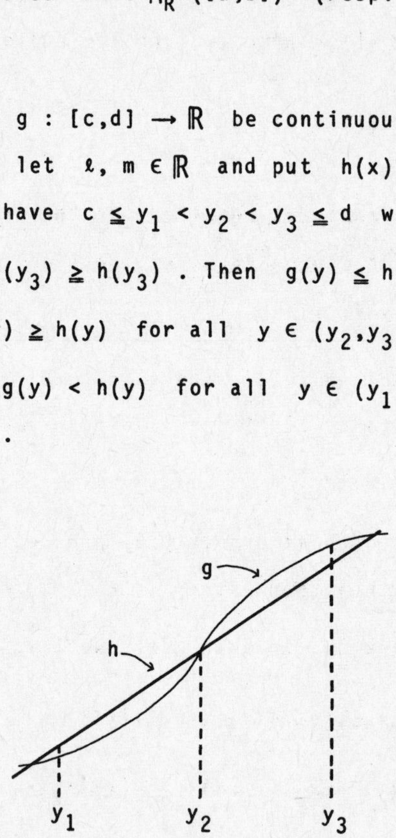

Proof Suppose first that g has property R^-; we can assume that $\ell > 0$ since the result is trivially true when $\ell \leq 0$. We will only show that $g(y) \geq h(y)$ for all $y \in (y_2,y_3)$; the proof of the other part is almost exactly the same. Let $y \in (y_2, y_3)$; if $g(y) \geq h(y_3)$ then we automatically have $g(y) > h(y)$ (since $\ell > 0$ implies h is increasing) and thus we can assume that $g(y) < h(y_3)$. We have

$$R(g(y_1),g(y_2),g(y),g(y_3)) = \frac{(g(y_3)-g(y_1))(g(y)-g(y_2))}{(g(y_3)-g(y))(g(y_2)-g(y_1))}$$

$$= \frac{(g(y_3)-g(y_1))(g(y)-h(y_2))}{(g(y_3)-g(y))(h(y_2)-g(y_1))} \leq \frac{(g(y_3)-h(y_1))(g(y)-h(y_2))}{(g(y_3)-g(y))(h(y_2)-h(y_1))}$$

(since $\dfrac{g(y_3)-z}{h(y_2)-z}$ is an increasing function of z in $(-\infty, h(y_2))$)

$$\leq \frac{(h(y_3)-h(y_1))(g(y)-h(y_2))}{(h(y_3)-g(y))(h(y_2)-h(y_1))}$$

(since $\dfrac{z-h(y_1)}{z-g(y)}$ is a decreasing function of z in $[h(y_3), +\infty)$).

But $R(g(y_1),g(y_2),g(y),g(y_3)) \geq R(y_1,y_2,y,y_3)$ and $R(y_1,y_2,y,y_3) = R(h(y_1),h(y_2),h(y),h(y_3))$. Thus

$$\frac{(h(y_3)-h(y_1))(g(y)-h(y_2))}{(h(y_3)-g(y))(h(y_2)-h(y_1))} \geq \frac{(h(y_3)-h(y_1))(h(y)-h(y_2))}{(h(y_3)-h(y))(h(y_2)-h(y_1))},$$

i.e. $\dfrac{g(y)-h(y_2)}{h(y_3)-g(y)} \geq \dfrac{h(y)-h(y_2)}{h(y_3)-h(y)}$. This implies that $g(y) \geq h(y)$, because $\dfrac{z-h(y_2)}{h(y_3)-z}$ is a strictly increasing function of z in $(h(y_2), h(y_3))$.

Now suppose g has property R ; again we can assume that $\ell > 0$. Let

$y \in (y_2,y_3)$ and choose $z_1 \in (y_1,y_2)$, $z_3 \in (y,y_3)$; then from the first part of the result (since g has property R´) we have $g(z_1) \leq h(z_1)$ and $g(z_3) \geq h(z_3)$. Repeating the first part of the proof gives us

$$R(g(z_1),g(y_2),g(y),g(z_3)) \leq \frac{(h(z_3)-h(z_1))(g(y)-h(y_2))}{(h(z_3)-g(y))(h(y_2)-h(z_1))} .$$

But $c < z_1 < y_2 < y < z_3 < d$ and g has property R ; thus $R(g(z_1),g(y_2),g(y),g(z_3)) > R(z_1,y_2,y,z_3)$, and so in the same way as above we get $\dfrac{g(y)-h(y_2)}{h(z_3)-g(y)} > \dfrac{h(y)-h(y_2)}{h(z_3)-h(y)}$. This implies that $g(y) > h(y)$. A similar proof shows that $g(y) < h(y)$ for all $y \in (y_1,y_2)$. ◻◻

Remark: At the end of the section (after the proof of Proposition 4.1) we will make a further study of the properties given in Proposition 4.6.

Proposition 4.6 in the case $\ell = 1$, $m = 0$ provides the key for proving Propositions 4.2, 4.4 and 4.5.

Proof of Proposition 4.5 The restriction of f to $[0,\varphi]$ has property R and $f(\gamma) = \gamma$, $f(\varphi) > \varphi$. Thus if $z < \gamma$ then Proposition 4.6 (with $\ell = 1$, $m = 0$) implies that $f(y) < y$ for all $y \in (z,\gamma)$. $f(y) > y$ for all $y \in [0,z)$ follows immediately from the fact that z is the smallest fixed point of f in $[0,\gamma]$. ◻◻

Proof of Proposition 4.4 Let $g \in M_R([a,b])$; then Proposition 4.6 (again with $\ell = 1$, $m = 0$) shows that g has at most three fixed points in each lap on which it is increasing. Thus if g has M turning points then $|Fix(g)| \leq 3(M+1)$, since g has $M+1$ laps and there is at most one fixed point in each lap on which g is decreasing. Now if $f \in M_R([a,b])$ has N turning points and $n \geq 1$ then by Proposition 2.1 we have

$$|T(f^n)| = |\{ x \in [a,b] : f^k(x) \in T(f) \text{ for some } 0 \leq k < n \}|$$

$$\leq N \sum_{k=0}^{n-1} (N+1)^k = (N+1)^n - 1$$

(where we have used the fact that $|f^{-1}(y)| \leq N+1$ for each $y \in [a,b]$). This, together with Lemma 4.2, gives us $|Fix(f^n)| \leq 3(N+1)^n$. ▯▯

Remark: Let $g \in M_R([a,b])$ have M turning points; a more careful application of Proposition 4.6 actually shows that $|Fix(g)| \leq 2+M$, and thus if $f \in M_R([a,b])$ has N turning points then for all $n \geq 1$ we have $|Fix(f^n)| \leq 2 + (N+1)^n$.

Proof of Proposition 4.2 Suppose $x \in Per(n,f)$ is trapped; then there exist $y < x < z$ and $\delta > 0$ such that f^{2n} is increasing on $[y-\delta, z+\delta]$ and $f^{2n}(y) \leq y$, $f^{2n}(z) \geq z$. By Lemma 4.2 the restriction of f^{2n} to $[y-\delta, z+\delta]$ has property R' and thus from Proposition 4.6 we have $f^{2n}(w) \leq w$ for all $w \in (y,x)$ and $f^{2n}(w) \geq w$ for all $w \in (x,z)$. Therefore x is an unstable fixed point of f^{2n} and hence also an unstable periodic point of f . Now let $[x]$ be free and non-exceptional; suppose x has period n and let $[u,v]$ be the largest interval containing x on which f^{2n} is increasing, thus $u < x < v$. From the proof of Proposition 2.6 (and replacing x by $f^j(x)$ for some $0 < j < n$ if necessary) we can assume that either $u \neq a$ and $f^{2n}(z) > z$ for all $z \in (u,x)$ or $v \neq b$ and $f^{2n}(z) < z$ for all $z \in (x,v)$. Without loss of generality suppose it is the latter that holds. v is then a turning point of f^{2n} and so by Proposition 2.1 there exists $w \in T(f)$ and $0 \leqq k < 2n$ with $f^k(v) = w$. If $f^{2n}(v) < v$ then exactly as in the proof of Proposition 2.7 we would have w is regular and $(w-\delta, w+\delta) \subset A([x],f)$ for some $\delta > 0$; thus we need only consider the case when $f^{2n}(v) = v$. Now the restriction of f^{2n} to $[u,v]$ has property R′ and so by Proposition 4.6 we have $f^{2n}(z) > z$ for all $z \in [u,x)$ (since $f^{2n}(x) = x$, $f^{2n}(v) = v$ and $f^{2n}(z) < z$ for all $z \in (x,v)$). If u is a turning point of f^{2n} then $f^j(u) = \overline{w}$ for some $\overline{w} \in T(f)$ and $0 \leqq j < 2n$, and as above this implies $\overline{w}$ is regular and $(\overline{w}-\delta, \overline{w}+\delta) \subset A([x],f)$ for some $\delta > 0$. Thus the only problem is when $u = a$.

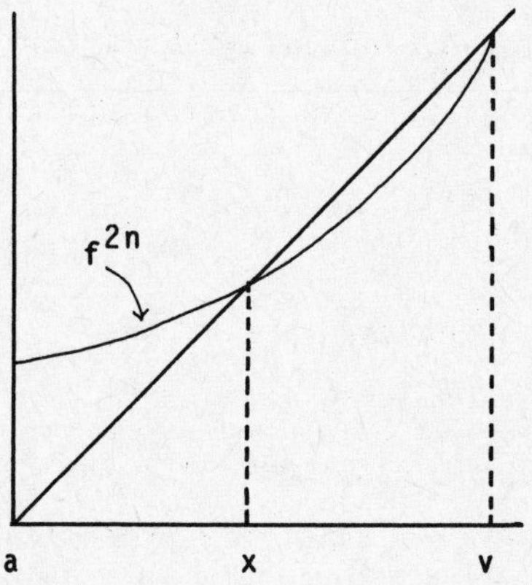

First suppose x is a fixed point of f (and hence $n = 1$); then, since $f(v)$ is also a fixed point of f^{2n} , we have $f(v) \geqq v$. This, together with the fact that $f(x) = x$ and f is monotone on $[a,v]$, gives us that f is actually increasing on $[a,x]$. It is now clear that (2.5) holds, and this is not possible because we are assuming $[x]$ is non-exceptional. Suppose then that $n > 1$. $f(v)$ is a fixed point of f^{2n} and $f(v) > v$. (We cannot have $f(v) = v$ because this would imply

x is critical.) Let $[u_1,v_1]$ be the largest interval containing $f(x)$ on which f^{2n} is increasing. We have $f(v) \in [u_1,v_1]$ (because f^m is monotone on $[x,v]$ for all $m \geq 0$) and $f(v)$ is a turning point of f^{2n} (since v is a turning point of f^{2n} and $f^{2n}(v) = v$); thus $f(v)$ is either u_1 or v_1. But $f(v)$ is a fixed point of f^{2n} and the restriction of f^{2n} to $[u_1,v_1]$ has property R$^-$; hence by Proposition 4.6 we have either $f^{2n}(z) > z$ for all $z \in [u_1,f(x))$ or $f^{2n}(z) < z$ for all $z \in (f(x),v_1]$. As before this will give us a regular $w \in T(f)$ and $\delta > 0$ with $(w-\delta,w+\delta) \subset A([x],f)$ except in the case when $f(v) = u_1$ and $v_1 = b$. If $f(v) = u_1$ and $v_1 = b$ then we consider the largest interval containing $f^2(x)$ on which f^{2n} is increasing; repeating the above argument we will then obtain our required regular element of $T(f)$ unless $f^2(x) = x$. Therefore the only case which remains to be dealt with is when $n = 2$ and the picture is as follows (where the functions drawn are parts of f^4):

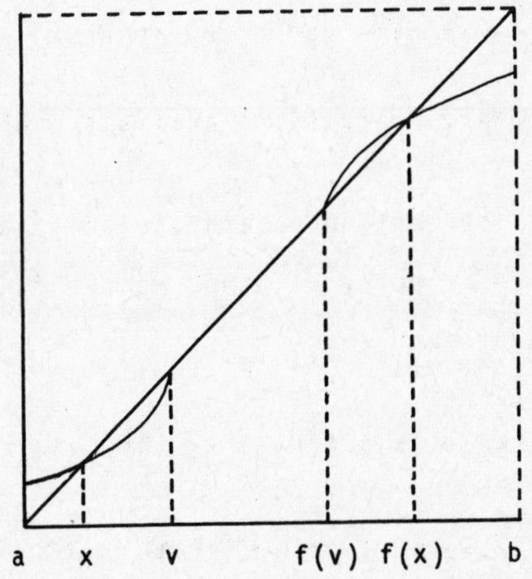

However, this case cannot occur: We have f^2 is increasing on $[a,v]$

(since $f^2(v)$ being a fixed point of f^{2n} implies $f^2(v) \geq v$), and hence $f^2(z) > z$ for all $z \in [a,x)$. Moreover, f is decreasing on $[a,v]$ because $f(v) < f(x)$; it then follows that f is also decreasing on $[f(v),b]$, and so (2.6) holds. But our assumptions have excluded this possibility. ▫▫

Proof of Proposition 4.1 We will only prove the part of the result which we have made use of; this is the part which says that if $(Sf)(x) < 0$ (resp. $(Sf)(x) \leq 0$) for all $x \in (a,b)$ then f has property R (resp. property R´). The proof of the converse is left to the reader. First we need a simple fact about linear fractional transformations.

Lemma 4.3 Let $y_k, z_k \in \mathbb{R}$, $k = 1, 2, 3$ with $y_1 < y_2 < y_3$ and $z_1 < z_2 < z_3$. Then there exist $\alpha, \beta, \gamma, \delta \in \mathbb{R}$ with $\gamma\beta-\alpha\delta \neq 0$ and $-\frac{\gamma}{\delta} \notin [y_1,y_3]$ such that if $T : \mathbb{R}-\{-\frac{\gamma}{\delta}\} \to \mathbb{R}$ is the linear fractional transformation given by $T(x) = \frac{\alpha+\beta x}{\gamma+\delta x}$ then $T(y_k) = z_k$ for $k = 1, 2, 3$. Moreover, T is strictly increasing on any interval not containing $-\frac{\gamma}{\delta}$.

Proof This is an elementary exercise in analysis. ▫▫

Now let $f : [a,b] \to \mathbb{R}$ be continuous and strictly increasing, and suppose f does not have property R ; then we can find $a < x_1 < x_2 < x_3 < x_4 < b$ such that

$$R(f(x_1),f(x_2),f(x_3),f(x_4)) \leq R(x_1,x_2,x_3,x_4) .$$

By Lemma 4.3 there exist $\alpha, \beta, \gamma, \delta \in \mathbb{R}$ with $\gamma\beta-\alpha\delta \neq 0$ and $-\frac{\gamma}{\delta} \notin [f(x_1),f(x_4)]$ and such that if T is the corresponding linear fractional transformation then $T(f(x_k)) = x_k$ for $k = 1, 2$ and 4 .

Choose $a \leqq c < x_1$ and $x_4 < d \leqq b$ so that $-\frac{\gamma}{\delta} \notin f([c,d])$ and define $g : [c,d] \to \mathbb{R}$ by $g(x) = T(f(x))$; then g is strictly increasing and $g(x_k) = x_k$ for $k = 1, 2$ and 4 . It is well-known (and in any case easily checked) that the cross-ratio is invariant under linear fractional transformations, and thus we have

$$R(f(x_1),f(x_2),f(x_3),f(x_4)) = R(g(x_1),g(x_2),g(x_3),g(x_4))$$

$$= R(x_1,x_2,g(x_3),x_4) .$$

Therefore $R(x_1,x_2,g(x_3),x_4) \leqq R(x_1,x_2,x_3,x_4)$ and so $\dfrac{g(x_3)-x_2}{x_4-g(x_3)} \leqq \dfrac{x_3-x_2}{x_4-x_3}$; this implies that $g(x_3) \leqq x_3$ (since $x_2 < g(x_3) < x_4$ and $\dfrac{z-x_2}{x_4-z}$ is a strictly increasing function of z in (x_2,x_4)).

Assume now that f has a third derivative in (a,b) and that $f'(x) \neq 0$ for all $x \in (a,b)$; then g also has these properties (in (c,d)). We have $g(x_k) = x_k$ for $k = 1, 2, 4$ and $g(x_3) \leqq x_3$; thus by the mean value theorem there exist $y_1 \in (x_1,x_2)$ with $g'(y_1) = 1$, $y_2 \in (x_2,x_3)$ with $g'(y_2) \leqq 1$ and $y_3 \in (x_3,x_4)$ with $g'(y_3) \geqq 1$. Therefore g' has a local minimum at some point $z \in (x_1,x_4)$ and for this z we have $g''(z) = 0$ and $g'''(z) \geqq 0$. Hence $(Sg)(z) \geqq 0$ (since $g'(z) > 0$). But $(Sf)(z) = (Sg)(z)$, because it is easily seen that the Schwarzian derivative is invariant under linear fractional transformations (and in fact it is for this reason that Schwarz originally introduced "his" derivative). We have thus shown that if $(Sf)(x) < 0$ for all $x \in (a,b)$ then f has property R , at least under the additional assumption that f is increasing. With minor alterations the proof also deals with the case when f is decreasing. (Alternatively, we could apply the above proof directly to the increasing function $\bar{f}$, where $\bar{f}(x) = f(a+b-x)$.) Finally, we must consider the case when we only have $(Sf)(x) \leq 0$ for all $x \in (a,b)$. For $\varepsilon > 0$ let $h_\varepsilon(x) = \varepsilon x^2 + x$ and put $g_\varepsilon = h_\varepsilon \circ f$. A direct calculation gives us

$$(Sg_\varepsilon)(x) = (Sh_\varepsilon)(f(x))[f'(x)]^2 + (Sf)(x) ,$$

and thus $(Sg_\varepsilon)(x) < 0$ because $(Sh_\varepsilon)(y) < 0$. Therefore if $\varepsilon > 0$ is small enough so that h_ε is strictly increasing on $f([a,b])$ then the first part of the proof implies that g_ε has property R . But if $a < x_1 < x_2 < x_3 < x_4 < b$ then

$$R(f(x_1),f(x_2),f(x_3),f(x_4)) = \lim_{\varepsilon\to 0} R(g_\varepsilon(x_1),g_\varepsilon(x_2),g_\varepsilon(x_3),g_\varepsilon(x_4)) ,$$

and hence f has property R´. ☐☐

We end this section by looking again at the properties given in Proposition 4.6. Let $g : [c,d] \to \mathbb{R}$ be continuous and strictly increasing; we say that g has *property Q* (resp. *property Q´*) if whenever $h : [c,d] \to \mathbb{R}$ is linear and we have $c \leq y_1 < y_2 < y_3 \leq d$ with $g(y_1) \leq h(y_1)$, $g(y_2) = h(y_2)$ and $g(y_3) \geq h(y_3)$ then $g(y) < h(y)$ for all $y \in (y_1,y_2)$ and $g(y) > h(y)$ for all $y \in (y_2,y_3)$ (resp. $g(y) \leq h(y)$ for all $y \in (y_1,y_2)$ and $g(y) \geq h(y)$ for all $y \in (y_2,y_3)$). Proposition 4.6 therefore says that property R´ (resp. property R) implies property Q´ (resp. property Q). The next result shows that functions which have property Q´ but not property Q are somewhat special.

Proposition 4.7 Let $g : [c,d] \to \mathbb{R}$ be continuous and strictly increasing; suppose g has property Q´ but not property Q . Then g is linear on some (non-trivial) sub-interval of $[c,d]$.

Proof Since g does not have property Q there exists a linear function $h : [c,d] \to \mathbb{R}$ and $c \leq y_1 < y_2 < y_3 \leq d$ with $g(y_1) \leq h(y_1)$, $g(y_2) = h(y_2)$ and $g(y_3) \geq h(y_3)$ such that either $g(z) \geq h(z)$ for some $z \in (y_1,y_2)$ or $g(z) \leq h(z)$ for some $z \in (y_2,y_3)$; without loss of generality assume it is the latter which holds. Now g has property Q´ and thus $g(y) \geq h(y)$ for all $y \in (y_2,y_3)$. In particular this implies that $g(z) = h(z)$. Again using the fact that g has property Q´ we thus also obtain $g(y) \leq h(y)$ for all $y \in (y_1,z)$, and hence $g(y) = h(y)$ for all $y \in [y_2,z]$. □□

Note that the converse of Proposition 4.7 is true in the sense that if g is linear on some non-trivial sub-interval of $[c,d]$ then it cannot have property Q . It is also worth noting that there exist non-trivial examples of functions having property R´ but not property Q , where by non-trivial we mean that the function is not linear on the whole of $[c,d]$. (Let $g : [c,d] \to \mathbb{R}$ be continuous with a third derivative in (c,d) and with $g'(x) > 0$ for all $x \in (c,d)$; suppose $(g')^{-1/2}$ is convex but that g' is constant on some non-trivial sub-interval of $[c,d]$. Then we have $(Sg)(x) \leq 0$ for all $x \in (c,d)$ and so by Proposition 4.1 g has property R´; however, g is linear on the sub-interval where g' is constant, and hence g does not have property Q.)

For functions having continuous derivatives we have the following simple characterizations of properties Q and Q´.

Proposition 4.8 Let $g : [c,d] \to \mathbb{R}$ be continuous, strictly increasing, and have a continuous derivative in (c,d). Then:

(1) g has property Q' if and only if $g'(\eta) \geq \min\{g'(a),g'(b)\}$ for all $c < a < \eta < b < d$;

(2) g has property Q if and only if g' has no local minimum in (c,d).

Note: If $f : [c,d] \to \mathbb{R}$ is continuous then f has no local minimum in (c,d) if and only if $f(\eta) > \min\{f(a),f(b)\}$ for all $c < a < \eta < b < d$.

Proof (1): Suppose g has property Q' and let $c < a < \eta < b < d$. Consider the line $h(x) = \alpha x+\beta$, where $\alpha = \min\left\{\frac{g(\eta)-g(a)}{\eta-a},\frac{g(b)-g(\eta)}{b-\eta}\right\}$ and $\beta = g(\eta) - \alpha\eta$. We thus have $g(a) \leq h(a)$, $g(\eta) = h(\eta)$ and $g(b) \geq h(b)$, and in fact either $g(a) = h(a)$ or $g(b) = h(b)$. Since g has property Q' we then have $g(x) \leq h(x)$ for all $x \in (a,\eta)$ and also $g(x) \geq h(x)$ for all $x \in (\eta,b)$. Therefore $g'(\eta) \geq g'(\xi)$, where

$$\xi = \begin{cases} a & \text{if } g(a) = h(a) , \\ b & \text{if } g(b) = g(b) ; \end{cases}$$ and thus $g'(\eta) \geq \min\{g'(a),g'(b)\}$.

Conversely, suppose that g does not have property Q'; then we can find a linear function $h : [c,d] \to \mathbb{R}$ and $c \leq y_1 < y_2 < y_3 \leq d$ with $g(y_1) \leq h(y_1)$, $g(y_2) = h(y_2)$ and $g(y_3) \geq h(y_3)$ and such that either $g(z) > h(z)$ for some $z \in (y_1,y_2)$ or $g(z) < h(z)$ for some $z \in (y_2,y_3)$. Without loss of generality assume the latter holds; by the mean value theorem we then have $g'(z_1) \geq \alpha$ for some $z_1 \in (y_1,y_2)$, $g'(z_2) < \alpha$ for some $z_2 \in (y_2,z)$ and $g'(z_3) > \alpha$ for some $z_3 \in (z,y_3)$, where α is the slope of h . This gives us $c < z_1 < z_2 < z_3 < d$ with $g'(z_2) < \min\{g'(z_1),g'(z_3)\}$.

(2): Suppose g has property Q but that g' has a local minimum at $\eta \in (c,d)$. Thus for some $\delta > 0$ we have $g'(x) \geq g'(\eta)$ for all $x \in (\eta-\delta,\eta+\delta)$. But g has property Q' and so by (1) we have $g'(\eta) \geq \min\{g'(a),g'(b)\}$ for all $c < a < \eta < b < d$. Therefore $g'(x) = g'(\eta)$ either for all $x \in (\eta-\delta,\eta)$ or for all $x \in (\eta,\eta+\delta)$. This cannot happen because it implies that g is linear on either $(\eta-\delta,\eta)$ or $(\eta,\eta+\delta)$. Conversely, suppose that g does not have property Q ; then exactly as in (1) we can find $c < z_1 < z_2 < z_3 < d$ with $g'(z_2) \leq \min\{g'(z_1),g'(z_3)\}$, and so g' has a local minimum somewhere in (z_1,z_3). ▫▫

Notes: The Schwarzian derivative was introduced in Schwarz (1868) and has since proved a powerful tool in the theory of conformal mapping (see, for example, Hille (1976), Chapter 10). The connection between negative Schwarzian derivatives and property R is given in Allwright (1978). He states (without proof) the following result, which is a bit stronger than Proposition 4.1:

Let $f : [a,b] \to \mathbb{R}$ be strictly monotone. Then f has property R' (resp. property R) if and only if f is continuously differentiable in (a,b) and $|f'|^{-1/2}$ is convex (resp. strictly convex).

The importance of a negative Schwarzian derivative in restricting the size of $P_s(f)$ was discovered by Singer, and Proposition 4.2 corresponds to Theorem 2.7 in Singer (1978). Theorem 4.3 is a version of Theorem 3.1 in Guckenheimer (1979). (Guckenheimer proves the harder result with "has Lebesgue measure 1 " replacing "contains a dense open subset of $[0,1]$ ", see Section 9.)

5. THE ITERATES OF FUNCTIONS IN S

In this section we analyse the possible types of behaviour which can be exhibited by the iterates of functions in S. Since in general things are rather complicated we start by describing what happens for functions in the set S_R^* consisting of those $f \in S_R$ for which

(5.1) f has a continuous second derivative in $(0,\varphi)\cup(\varphi,1)$ and $f'(z) \neq 0$ for all $z \in (0,\varphi)\cup(\varphi,1)$;

(5.2) $\limsup_{z\uparrow\varphi} \left|\frac{f'(z)}{f'(\underline{z})}\right| < +\infty$ and $\liminf_{z\uparrow\varphi} \left|\frac{f'(z)}{f'(\underline{z})}\right| > 0$, where $\underline{z} \neq z$ is such that $f(\underline{z}) = f(z)$.

It is easy to check that (5.2) holds if

(5.3) f has a continuous second derivative in $(0,1)$, $f'(z) \neq 0$ for all $z \in (0,\varphi)\cup(\varphi,1)$ and $f''(\varphi) < 0$,

(since in this case we actually have $\lim_{z\uparrow\varphi} \left|\frac{f'(z)}{f'(\underline{z})}\right| = 1$). Thus S_R^* contains all functions in S_R for which (5.3) holds, and so in particular contains the functions in the families $f_\mu(x) = \mu x(1-x)$, $0 < \mu \leq 4$, $f_\mu(x) = \sin(\mu x)$, $\frac{\pi}{2} < \mu \leq \pi$, and $f_\mu(x) = \mu x \exp(1-\mu x)$, $\mu > 1$.

In order to describe the iterates of functions in S_R^* we need a couple of definitions. Recall that $f \in S$ is *orbit continuous* at $x \in [0,1]$ if for each $\varepsilon > 0$ there exists $\delta > 0$ such that $|f^n((x-\delta,x+\delta))| < \varepsilon$ for all $n \geq 0$; $O(f)$ denotes the set of points in $[0,1]$ at which f is orbit continuous. (*Note:* By $(x-\delta,x+\delta)$ we of course mean $\{ y \in [0,1] : x-\delta < y < x+\delta \}$.) Now for $f \in S$ and $\varepsilon > 0$ let

$$\Sigma_\varepsilon(f) = \{ x \in [0,1] : \text{for each } \delta > 0 \text{ there exists } n \geq 0 \text{ such that } |f^k((x-\delta,x+\delta))| \geq \varepsilon \text{ for all } k \geq n \} ;$$

clearly for each $\varepsilon > 0$ we have $\Sigma_\varepsilon(f) \cap O(f) = \phi$. We say that $f \in S$ has *sensitive dependence to initial conditions* if there exists $\varepsilon > 0$ such that $\Sigma_\varepsilon(f) = [0,1]$. At first sight this definition might seem too weak to be of any practical use. However, it does imply the kind of sensitive dependence described in the Introduction: For $f \in S$, $\varepsilon > 0$ and $x \in [0,1]$ let

$$\Omega_\varepsilon(x,f) = \{ y \in [0,1] : |f^n(y)-f^n(x)| > \varepsilon \text{ for some } n \geq 0 \} ;$$

$$\Omega^*_\varepsilon(x,f) = \{ y \in [0,1] : |f^n(y)-f^n(x)| > \varepsilon \text{ for infinitely many } n \geq 0 \} ;$$

the following then holds:

Lemma 5.1 Let $f \in S$, $0 < 2\eta < \varepsilon$ and suppose $\Sigma_\varepsilon(f) = [0,1]$. Then for each $x \in [0,1]$ we have $\Omega_\eta(x,f)$ (resp. $\Omega^*_\eta(x,f)$) is a dense open subset (resp. a residual subset) of $[0,1]$.

Proof For $m \geq 0$ let

$$\Omega^m_\eta(x,f) = \{ y \in [0,1] : |f^n(y)-f^n(x)| > \eta \text{ for some } n \geq m \} ;$$

then $\Omega^m_\eta(x,f)$ is open and we have $\Omega_\eta(x,f) = \Omega^0_\eta(x,f)$ and $\Omega^*_\eta(x,f) = \bigcap_{m\geq 0} \Omega^m_\eta(x,f)$. It is thus enough to show that each $\Omega^m_\eta(x,f)$ is dense. Let $J \subset [0,1]$ be a non-trivial interval; then there exists $n \geq 0$ such that $|f^k(J)| \geq \varepsilon$ for all $k \geq n$. In particular there exists $k \geq m$ with $|f^k(J)| \geq \varepsilon > 2\eta$, and so $\Omega_\eta^m(x,f) \cap J \neq \phi$. Hence $\Omega_\eta^m(x,f)$ is dense in $[0,1]$. □□

Lemma 5.1 says that if f has sensitive dependence to initial conditions then there exists $\varepsilon > 0$ such that $\Omega_\varepsilon^*(x,f)$ is a residual subset of $[0,1]$ for each $x \in [0,1]$; this means that for each $x \in [0,1]$ and for each "typical" point y chosen in any neighbourhood of x we have $|f^n(y)-f^n(x)| > \varepsilon$ for infinitely many $n \geq 0$.

Let $f \in \mathcal{S}$ and let $\{I^{(n)}\}_{n\geq 1}$ be a decreasing sequence of closed subsets of $[0,1]$; we say that $(\{I^{(n)}\}_{n\geq 1},f)$ is an *infinite register shift* if for each $n \geq 1$ there exists an integer $m_n \geq 2$ such that, putting $q_n = \prod_{k=1}^{n} m_k$, we have

(5.4) for each $n \geq 1$ $I^{(n)}$ is the disjoint union of q_n non-trivial closed intervals $I_0^{(n)},\ldots,I_{q_n-1}^{(n)}$ such that $f(I_j^{(n)}) = I_{j+1}^{(n)}$ for $j = 0,\ldots,q_n-1$ (where $I_{q_n}^{(n)} = I_0^{(n)}$) .

It is easy to see that a direct consequence of (5.4) is:

(5.5) each of the q_n intervals making up $I^{(n)}$ contains m_{n+1} of the intervals making up $I^{(n+1)}$.

(In fact, if we label the intervals so that (5.4) holds and also so that $I_0^{(n+1)} \subset I_0^{(n)}$ then $I_k^{(n+1)} \subset I_j^{(n)}$ if and only if $k \equiv j \pmod{q_n}$.)

We say that an infinite register shift $(\{I^{(n)}\}_{n\geq 1},f)$ is *proper* if $I^{(\infty)} = \bigcap_{n\geq 1} I^{(n)}$ is a nowhere dense subset of $[0,1]$. In this case (because of (5.5)) $I^{(\infty)}$ is a generalized Cantor-like set.

Proposition 5.1 Let $f \in S$ and $(\{I^{(n)}\}_{n\geq 1}, f)$ be a proper infinite register shift; put $I^{(\infty)} = \bigcap_{n\geq 1} I^{(n)}$. Then f maps $I^{(\infty)}$ homeomorphically back onto itself. Furthermore, for each $x \in I^{(\infty)}$ the orbit $\{f^n(x)\}_{n\geq 0}$ is dense in $I^{(\infty)}$.

Proof We leave the reader to check that f maps $I^{(\infty)}$ bijectively onto itself. This then implies that f maps $I^{(\infty)}$ homeomorphically onto itself (since a continuous bijection of a compact Hausdorff space onto itself is automatically a homeomorphism). Now take $x, y \in I^{(\infty)}$; for each $n \geq 1$ there exists $0 \leq j < q_n$ and $k \geq 0$ such that both y and $f^k(x)$ are in $I_j^{(n)}$, and hence for each $n \geq 1$ there exists $k \geq 0$ such that $|f^k(x)-y| \leq \max_{0\leq j<q_n} |I_j^{(n)}|$. But $\lim_{n\to\infty} \max_{0\leq j<q_n} |I_j^{(n)}| = 0$ because $I^{(\infty)}$ is nowhere dense, and this shows that the orbit $\{f^k(x)\}_{k\geq 0}$ is dense in $I^{(\infty)}$. ▫▫

We are now in a position to state the main classification result for functions in S_R^*.

Theorem 5.1 Let $f \in S_R^*$ with $f(z) > z$ for all $z \in (0,\varphi]$. Then exactly one of the following statements holds:

(5.6) $|P_s(f)| = 1$, and if $[x] \in P_s(f)$ then $A([x],f)$ contains a dense open subset of $[0,1]$.

(5.7) $|P_s(f)| = 0$ and f has sensitive dependence to initial conditions.

(5.8) $|P_s(f)| = 0$ and there exists a decreasing sequence $\{I^{(n)}\}_{n \geq 1}$ of closed subsets of $[0,1]$, each containing φ in its interior, such that $(\{I^{(n)}\}_{n \geq 1}, f)$ is a proper infinite register shift; moreover, for each $n \geq 1$ $\{ x \in [0,1] : f^k(x) \in \mathrm{int}(I^{(n)})$ for some $k \geq 0 \}$ is a dense open subset of $[0,1]$.

Proof This will be a special case of Theorem 5.2. ▫▫

Remark: The assumption that $f(z) > z$ for all $z \in (0,\varphi]$ was only made to simplify the statement of the result. We leave the reader to deduce from Theorem 5.2 what happens without this assumption.

Theorem 5.1 gives a precise formulation of the three types of behaviour mentioned in the Introduction, and shows that if $f \in S_R^*$ with $f(z) > z$ for all $z \in (0,\varphi]$ then these three types completely classify the possible behaviour which can be exhibited by the iterates of f. In Section 10 we will show that each of these types actually occurs. In fact we will see that any "reasonable" one-parameter family of functions from S_R^* (such as $f_\mu(x) = \mu x(1-x)$, $0 < \mu \leq 4$, $f_\mu(x) = \sin(\mu x)$, $\frac{\pi}{2} < \mu \leq \pi$, and $f_\mu(x) = \mu x \exp(1-\mu x)$, $\mu > 1$) contains infinitely many functions of each type.

In Section 9 we give an "almost all" version of Theorem 5.1. We will prove that if $f \in S_R^*$ with $f(z) > z$ for all $z \in (0,\varphi]$ then:

(i) If (5.6) holds and $[x] \in P_s(f)$ is stable then $\lambda(A([x],f)) = 1$, (where λ denotes Lebesgue measure defined on the Borel subsets of $[0,1]$).

(ii) If (5.8) holds and $(\{I^{(n)}\}_{n \geq 1}, f)$ is the corresponding proper infinite register shift then for each $n \geq 1$

$$\lambda(\{ x \in [0,1] : f^k(x) \in \mathrm{int}(I^{(n)}) \text{ for some } k \geq 0 \}) = 1 .$$

Suppose that (5.8) holds and let $G = \bigcap_{n\geq 1} G_n$, where $G_n = \{ x \in [0,1] : f^k(x) \in \mathrm{int}(I^{(n)})$ for some $k \geqq 0 \}$. Then G is a residual subset of $[0,1]$ and the points in G are attracted to the Cantor-like set $I^{(\infty)} = \bigcap_{n\geq 1} I^{(n)}$: If $x \in G$ then for each $n \geqq 1$ there exists $k \geqq 0$ such that $f^j(x) \in I^{(n)}$ for all $j \geqq k$. In particular $\lim_{n\to\infty} \min_{z\in I^{(\infty)}} |f^n(x)-z| = 0$ for each $x \in G$. We also have:

Proposition 5.2 Let $f \in S$ and suppose that (5.8) holds. Then $O(f)$ is a residual subset of $[0,1]$ and $\Sigma_\varepsilon(f)$ is nowhere dense for each $\varepsilon > 0$.

Proof Let $\varepsilon_n = \max_{0\leqq j<q_n} I_j^{(n)}$; since $I^{(\infty)}$ is nowhere dense we have $\lim_{n\to\infty} \varepsilon_n = 0$. Now let $n \geqq 1$ and $x \in G_n$, so $f^k(x) \in \mathrm{int}(I^{(n)})$ for some $k \geqq 0$. If $\delta > 0$ is small enough then for any $y \in (x-\delta,x+\delta)$ we have $|f^j(y)-f^j(x)| < \varepsilon_n$ for $j = 0,\ldots,k$ and also $f^k(x)$ and $f^k(y)$ are both in the same component of $I^{(n)}$. By (5.4) we then have $|f^j(y)-f^j(x)| < \varepsilon_n$ for all $j \geqq 0$. This shows that $\bigcap_{n\geqq 1} G_n \subset O(f)$, and thus $O(f)$ is residual. It also shows that if $\varepsilon > \varepsilon_n$ then $\Sigma_\varepsilon(f) \subset [0,1] - G_n$, and hence $\Sigma_\varepsilon(f)$ is nowhere dense for each $\varepsilon > 0$.

□□

Putting together Theorem 4.3, Theorem 5.1 and Proposition 5.2 we see that if $f \in S_R^*$ with $f(z) > z$ for all $z \in (0,\varphi]$ then the following holds for the sets $A(f)$, $O(f)$ and $\Sigma_\varepsilon(f)$ (where we recall that $A(f) = \{ y \in [0,1] : y \in A([x],f)$ for some $[x] \in P(f) \}$

$$= \{ x \in [0,1] : \lim_{m\to\infty} f^{nm}(x) \text{ exists for some } n \geqq 1 \}) :$$

	(5.6) holds	(5.7) holds	(5.8) holds
$A(f)$	contains a dense open subset of $[0,1]$	is countable	is countable
$O(f)$	contains a dense open subset of $[0,1]$	is empty	is residual
$\Sigma_\varepsilon(f)$	is nowhere dense for each $\varepsilon > 0$	equals $[0,1]$ for some $\varepsilon > 0$	is nowhere dense for each $\varepsilon > 0$

Which of the three types of behaviour occurs for a given function f can be determined from just knowing what happens in any neighbourhood of φ. More precisely, we have:

Proposition 5.3 Let $f \in S_R^*$ with $f(z) > z$ for all $z \in (0,\varphi]$. Then

(1) (5.6) holds if and only if φ is either regular or a stable periodic point; moreover, this happens if and only if $\varphi \in A(f) \cap O(f)$.

(2) (5.7) holds if and only if $\varphi \in \Sigma_\varepsilon(f)$ for some $\varepsilon > 0$; moreover, this happens if and only if $\Omega_\varepsilon^*(\varphi,f)$ is residual for some $\varepsilon > 0$.

(3) (5.8) holds if and only if $\varphi \in O(f) - A(f)$.

Proof The first part of (1) follows from Theorem 4.3, and this implies also that if (5.6) holds then $\varphi \in A(f) \cap O(f)$. If (5.7) holds then clearly $\varphi \in \Sigma_\varepsilon(f)$ and by Lemma 5.1 $\Omega_\varepsilon^*(\varphi,f)$ is residual for some $\varepsilon > 0$. Suppose that (5.8) holds; the proof of Proposition 5.2 shows that

$\varphi \in O(f)$. We also have $\varphi \notin A(f)$, because if $\varphi \in A([x],f)$ for some x then we would have $x \in I^{(\infty)}$, and this is not possible since $I^{(\infty)}$ contains no periodic points. (Note that any periodic point in $I^{(n)}$ has a period which is a multiple of q_n , and $\lim_{n\to\infty} q_n = +\infty$.) Thus $\varphi \in O(f) - A(f)$. The converses of these statements follow immediately because exactly one of (5.6), (5.7) and (5.8) holds, $O(f) \cap \Sigma_\varepsilon(f) = \phi$ for all $\varepsilon > 0$, and because if $\varphi \in O(f)$ then for each $\varepsilon > 0$ there exists $\delta > 0$ with $(\varphi-\delta,\varphi+\delta) \cap \Omega^*_\varepsilon(\varphi,f) = \phi$. ⊞

We can in fact improve on Proposition 5.3 in that it is almost possible to determine which of the three types of behaviour occurs for a given function f from just knowing the orbit $\{f^n(\varphi)\}_{n\geq 0}$. (The one case where this will not always be possible is when φ is periodic.) Let $\mathcal{W} = \{ \{\theta_n\}_{n\geq 0} : \theta_n \in \{-1,0,1\}$ for each $n \geq 0 \}$ and for $f \in \mathcal{S}$ define $k_f = \{k_f(n)\}_{n\geq 0} \in \mathcal{W}$ by

$$k_f(n) = \begin{cases} -1 & \text{if } f^n(\varphi) < \varphi , \\ 0 & \text{if } f^n(\varphi) = \varphi , \\ 1 & \text{if } f^n(\varphi) > \varphi . \end{cases}$$

k_f is called the *kneading sequence* of f and in Section 8 we show that if $f \in \mathcal{S}^*_R$ with $f(z) > z$ for all $z \in (0,\varphi]$ is such that φ is not periodic then we can determine which one of (5.6), (5.7) and (5.8) holds from just knowing k_f . Of course, the kneading sequence is really nothing more than a special case of the "codes" introduced at the end of Section 2 (and in terms of which we defined whether or not φ is regular). Note that if $f \in \mathcal{S}$ then φ is either regular or periodic if and only if $\{k_f(n+1)\}_{n\geq 0}$ is periodic.

The result in Section 8 concerning kneading sequences will follow from an analysis of the proof of Theorem 5.1; in this section we give a somewhat weaker result (Proposition 5.4) which follows directly from the statement of Theorem 5.1. As a preparation for this we introduce a property which only depends on $\{f^n(\varphi)\}_{n\geq 0}$ and which holds if and only if (5.8) does. Let $f \in \mathcal{S}$ and $x \in [0,1]$; we call x *pseudo-periodic* if for each $\varepsilon > 0$ there exists $n \geq 1$ such that

(5.9) $|f^j(x)-f^k(x)| < \varepsilon$ whenever $j = k \pmod n$,

(5.10) x does not lie between $f^j(x)$ and $f^k(x)$ if $j = k \pmod n$ and $j \neq 0 \pmod n$.

Clearly any periodic point x is pseudo-periodic (just take n to be the period of x for each $\varepsilon > 0$).

Lemma 5.2 Let $f \in \mathcal{S}$ and $(\{I^{(n)}\}_{n\geq 1}, f)$ be a proper infinite register shift. Then each point in $I^{(\infty)} = \bigcap_{n\geq 1} I^{(n)}$ is pseudo-periodic but not periodic.

Proof We have already noted that no point in $I^{(\infty)}$ can be periodic. Let $\varepsilon_m = \max_{0\leq j<q_m} |I_j^{(m)}|$; so $\lim_{m\to\infty} \varepsilon_m = 0$. Now given $\varepsilon > 0$ we choose $m \geq 1$ so that $\varepsilon_m < \varepsilon$ and then take $n = q_m$. If $x \in I^{(\infty)}$ and $j = k \pmod n$ then $f^j(x)$ and $f^k(x)$ are in the same component of $I^{(m)}$ and thus $|f^j(x)-f^k(x)| \leq \varepsilon_m < \varepsilon$; if also $j \neq 0 \pmod n$ then x is not in this component and hence x cannot lie between $f^j(x)$ and $f^k(x)$. Therefore x is pseudo-periodic. ☐☐

Lemma 5.3 Let $f \in \mathcal{S}$ and suppose φ is pseudo-periodic but not periodic. Then $\varphi \in O(f)$.

Proof Let $\varepsilon > 0$ and choose $n \geq 1$ so that (5.9) and (5.10) hold.

Since φ is not periodic we have $f^n(\varphi) \neq \varphi$; without loss of generality assume that $f^n(\varphi) > \varphi$. Now using (5.10) it is easy to see that for each $k \geq 0$ the end-points of the interval $f^k([\varphi, f^n(\varphi)])$ are of the form $f^j(\varphi)$ with $j = k \pmod n$. Thus by (5.9) we have $|f^k([\varphi, f^n(\varphi)])| < \varepsilon$ for all $k \geq 0$, and hence $|f^k(\varphi-\delta, \varphi+\delta)| < \varepsilon$ for all $k \geq 0$ provided $0 < \delta < \varepsilon$ is chosen so that $(\varphi-\delta, \varphi+\delta) \subset f^{-1}([f^{n+1}(\varphi), f(\varphi)])$. Therefore $\varphi \in O(f)$. ▫▫

Proposition 5.4 Let $f \in S_R^*$ with $f(z) > z$ for all $z \in (0, \varphi]$. If φ is not periodic then

(1) (5.6) holds if and only if φ is regular;

(2) (5.7) holds if and only if φ is neither regular nor pseudo-periodic;

(3) (5.8) holds if and only if φ is pseudo-periodic.

If φ is periodic then

(4) (5.6) holds if and only if φ is stable;

(5) (5.7) holds if and only if φ is unstable.

Proof (1) and (4) follow from Proposition 5.3(1).

(3): If (5.8) holds then by Lemma 5.2 φ is pseudo-periodic. Conversely, if φ is pseudo-periodic (but not periodic) then by Lemma 5.3 we have $\varphi \in O(f) - A(f)$, and hence by Proposition 5.3(3) (5.8) holds.

(2): This now follows because exactly one of (5.6), (5.7) and (5.8) holds.

(5): This is clear. ▫▫

Note that we can determine whether φ is regular or pseudo-periodic from just knowing the orbit $\{f^n(\varphi)\}_{n\geq 0}$.

Proposition 5.4 can be simplified if we make the additional assumption that f has a continuous derivative in $[0,1]$. In this case φ cannot be an unstable periodic point, and this gives us the following result:

Let $f \in S_R^*$ with $f(z) > z$ for all $z \in (0,\varphi]$ and suppose f has a continuous derivative in $[0,1]$. Then

(1) (5.6) holds if and only if φ is either regular or periodic;

(2) (5.7) holds if and only if φ is neither regular nor pseudo-periodic;

(3) (5.8) holds if and only if φ is pseudo-periodic but not periodic.

We now move on to consider what happens for a general function in S . First we need some more definitions. Let $f \in S$; recall that a closed, non-trivial interval $J \subset [0,1]$ is called a *sink* of f if for some $n \geq 1$ we have f^n is monotone on J and $f^n(J) \subset J$; if J is a sink then of course f^m is monotone on J for all $m \geq 0$. We call a sink *maximal* if it is not a proper subset of any other sink.

Lemma 5.4 Let $f \in S$; then each sink of f is contained in a maximal sink.

Proof Let $J = [a,b]$ be a sink of f and let n be the smallest positive integer with $f^n(J) \subset J$. Put

$$m = \begin{cases} n & \text{if } f^n \text{ is increasing on } J , \\ 2n & \text{if } f^n \text{ is decreasing on } J ; \end{cases}$$

thus f^m is increasing on J and $f^m(J) \subset J$. Let $[u,v]$ be the largest interval containing J on which f^m is increasing and define

$$\overline{u} = \begin{cases} \text{the smallest fixed point of } f^m \text{ in } [u,v] & \text{if } f^m(u) < u , \\ u & \text{if } f^m(u) \geqq u , \end{cases}$$

$$\overline{v} = \begin{cases} \text{the largest fixed point of } f^m \text{ in } [u,v] & \text{if } f^m(v) > v , \\ v & \text{if } f^m(v) \leqq v . \end{cases}$$

Clearly $f^m(\overline{u}) \geqq \overline{u}$ and $f^m(\overline{v}) \leqq \overline{v}$, and thus $f^m([\overline{u},\overline{v}]) \subset [\overline{u},\overline{v}]$. We also have $[a,b] \subset [\overline{u},\overline{v}]$, because $f^m(a) \geqq a$ and $f^m(b) \leqq b$, and so in particular $\overline{u} < \overline{v}$. Hence $[\overline{u},\overline{v}]$ is a sink.

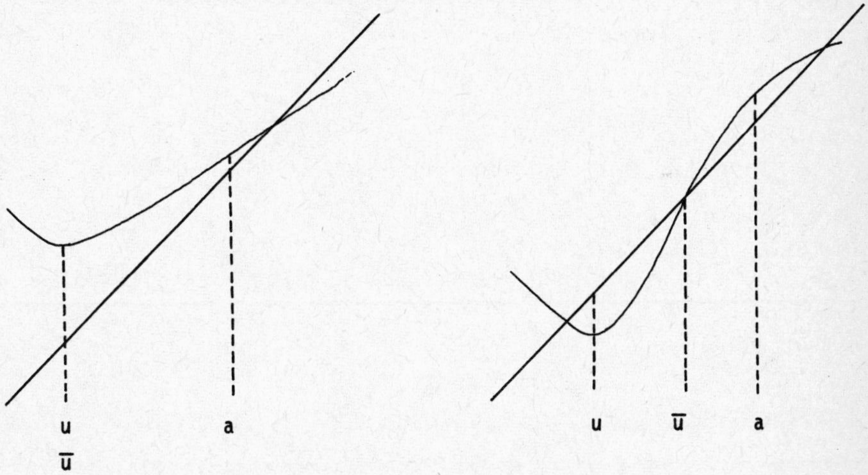

It is now left as a simple exercise for the reader to show that $[\overline{u},\overline{v}]$ is maximal. ◻◻

Remarks: (1) The integer m defined in the above proof is of course the smallest positive integer such that both f^m is increasing on J and $f^m(J) \subset J$. Note however that if m is replaced by any other positive

integer having these two properties then $\overline{u}$ and $\overline{v}$ remain the same.
(2) If J_1 and J_2 are sinks of f with $J_1 \cap J_2 \neq \phi$ then it is easily checked that $J_1 \cup J_2$ is also a sink. It thus follows that two maximal sinks are either disjoint or equal.

Our analysis of the iterates of a function $f \in S$ involves a classification of the maximal sinks of f. Let $J = [a,b]$ be a sink of f and as in the proof of Lemma 5.4 let m be the smallest positive integer such that both f^m is increasing on J and $f^m(J) \subset J$; let $[u,v]$ be the largest interval containing J on which f^m is increasing. We call J a *trap* of f if $u < a < b < v$ and $f^m(z) < z$ for all $z \in [u,a)$, $f^m(z) > z$ for all $z \in (b,v]$. (Note that a and b are then fixed points of f^m, and it is easy to see that they are both trapped periodic points of f.)

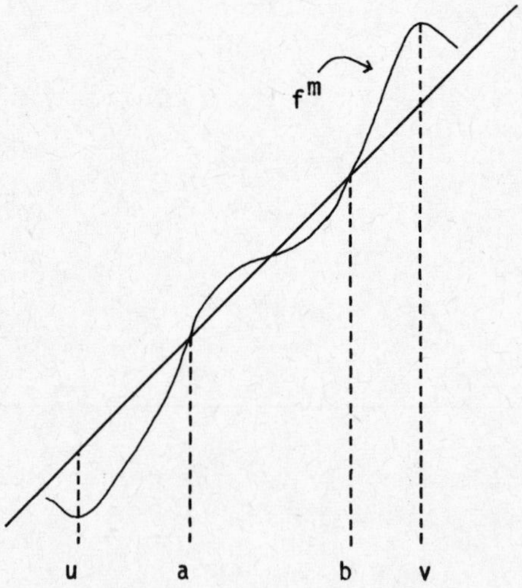

Remark: In the definition of a trap it would not make any difference if we replaced m by any positive integer k such that both f^k is increasing on J and $f^k(J) \subset J$ (cf. Remark (1) following Lemma 5.4). Clearly a trap is a maximal sink. Note that if J is a trap then so is $f^k(J)$ for each $k \geq 0$.

Proposition 5.5 If $f \in S_R$. then f has no traps.

Proof This follows immediately from Proposition 4.6. ◻◻

We call $f \in S$ *central* if there exists a sink J of f having φ as an end-point. If there is a sink having φ as left-hand end-point (resp. right-hand end-point) then there can be no sink having φ as right-hand end-point (resp. left-hand end-point). Thus if f is central then there exists a unique maximal sink K having φ as an end-point; we call K the *central sink* of f.

Remark: If $f \in S$ with $f(\varphi) \leq \varphi$ then f is central and $[0,\varphi]$ is the central sink.

Proposition 5.6 Let $f \in S_R$ with $f(\varphi) > \varphi$. Then f is central if and only if $P_s(f)$ contains an element which is not a fixed point of f in $[0,\varphi]$. In particular, if $f \in S_R$ with $f(z) > z$ for all $z \in (0,\varphi]$ then f is central if and only if $|P_s(f)| = 1$.

Proof First note that if $f \in S_R$ then any sink of f contains a periodic point x with $[x] \in P_s(f)$. (This follows because by Proposition 4.4 f^m has only finitely many fixed points for each $m \geq 1$.) If f is central and K is the central sink of f then (since $f(\varphi) > \varphi$) K cannot contain a fixed point of f in $[0,\varphi]$. Thus if f is central then $P_s(f)$ contains an element which is not a fixed point of f in $[0,\varphi]$. Conversely, suppose that $P_s(f)$ does contain an element which is not a fixed point of f in $[0,\varphi]$. Theorem 4.2 then implies that φ is either regular or a stable periodic point and in both cases it is easy to see that there exists a sink having φ as an end-point; i.e. f is central. ⧉

Let $f \in S$ with $f(\varphi) > \varphi$; as in Section 3 we put

$$\gamma = \begin{cases} \text{the largest fixed point of } f \text{ in } [0,\varphi] & \text{if } f \text{ has a fixed point in } [0,\varphi] , \\ 0 & \text{otherwise,} \end{cases}$$

$$\bar{\gamma} = \begin{cases} \text{the unique point in } [\varphi,1] \text{ with } f(\bar{\gamma}) = \gamma & \text{if } f(1) \leqq \gamma , \\ 1 & \text{otherwise.} \end{cases}$$

Note that if $f(\varphi) > \varphi$ and $\gamma > 0$ then $[0,\gamma]$ is a maximal sink.

Proposition 5.7 Let $f \in S$ with $f(\varphi) > \varphi$ and let $J \subset [\gamma,\bar{\gamma}]$ be a maximal sink. If J is not a trap then f is central and we have $f^n(J) \subset K$ for some $n \geqq 0$ (where K is the central sink).

Proof Let $J = [a,b]$ and let m be the smallest positive integer such that both $f^m(J) \subset J$ and f^m is increasing on J ; let $[u,v]$ be the largest interval containing J on which f^m is increasing. Since J is maximal but not a trap we have either $u = a$ or $b = v$. Suppose first that either $u = a > 0$ or $b = v < 1$; then by Proposition 2.1 there exists $n \geqq 0$ such that φ is an end-point of $f^n(J)$. But $f^n(J)$ is a sink, hence f is central and of course $f^n(J) \subset K$. Suppose next that $b = 1$; we then have $f(\varphi) > a$ and $f^m([a,1]) \subset [a,f(\varphi)]$ (since $f([0,1]) \subset [0,f(\varphi)]$). Let $[c,d] = f^{-1}([a,f(\varphi)])$, so $c < \varphi < d$ and it is easily checked that $f^m([c,d]) \subset [c,d]$. It thus follows that either $[c,\varphi]$ or $[\varphi,d]$ is a sink, and hence f is central; also we have $f^{m-1}([a,1]) \subset K$ because $f^{m-1}([a,1]) \subset [c,d]$. Finally, suppose that

$a = 0$; in this case we must have $f(z) > z$ for all $z \in [0,\varphi]$ and so $f^n([0,1]) \subset [f^2(\varphi),1]$ for all large enough n . Thus $f^2(\varphi) < b$ and if $[c,d] = f^{-1}([\varphi,1] \cap f^{-1}([f^2(\varphi),b]))$ then again either $[c,\varphi]$ or $[\varphi,d]$ is a sink. Therefore f is central and we have $f^n([0,b]) \subset K$ for some $n \geq 0$ because we can find $n \geq 0$ with $f^n([0,b]) \subset [c,d]$. □□

Let $f \in \mathcal{S}$ and J be a sink of f . If $f(\varphi) \leq \varphi$ then clearly $J \subset [0,\varphi]$; if $f(\varphi) > \varphi$ then Proposition 5.7 tells us that exactly one of the following holds: *(i)* $J \subset [0,\gamma]$, *(ii)* J is contained in a trap, *(iii)* f is central and $f^n(J) \subset K$ for some $n \geq 0$.

We need to introduce one more special kind of interval: Let $f \in \mathcal{S}$ and $J \subset [0,1]$ be a non-trivial closed interval; we call J a *homterval* of f if $\varphi \notin f^n(J)$ for all $n \geq 0$ and $f^n(J)$ is not contained in any sink for each $n \geq 0$.

Proposition 5.8 (1) If J is a homterval of $f \in \mathcal{S}$ then the intervals $\{f^n(J)\}_{n \geq 0}$ are disjoint.

(2) Let $f \in \mathcal{S}$ and $J \subset [0,1]$ be a non-trivial closed interval such that $\varphi \notin f^n(J)$ for all $n \geq 0$; let $x \in J$. Then J is a homterval if and only if $x \notin A(f)$.

Proof (1) Let J be a non-trivial closed interval with $\varphi \notin f^n(J)$ for all $n \geq 0$ and suppose that $f^j(J) \cap f^k(J) \neq \phi$ for some $0 \leq j < k$. Then $f^{j+n}(J) \cap f^{k+n}(J) \neq \phi$ for all $n \geq 0$ and so putting $I_m = f^{j+m(k-j)}(J)$ we have $I_m \cap I_{m+1} \neq \phi$ for each $m \geq 0$. Let $I = \bigcup_{m \geq 0} I_m$; thus I is an interval, $\varphi \notin f^n(I)$ for all $n \geq 0$ and $f^{k-j}(I) \subset I$. Therefore $\overline{I}$ is a sink and $f^j(J) \subset \overline{I}$, i.e. J is not a homterval.

(2) If $f^k(J)$ is contained in a sink for some $k \geq 0$ then clearly

$x \in A(f)$. Conversely, suppose that $x \in A(f)$; there thus exists $m \geq 1$ and $z \in [0,1]$ with $\lim_{n\to\infty} f^{nm}(x) = z$, and z must be a fixed point of f^m . If $[z] \in P_s(f)$ then for some $\delta > 0$ either $[z,z+\delta]$ or $[z-\delta,z]$ is a sink containing all but finitely many members of the sequence $\{f^{2nm}(J)\}_{n\geq 0}$. If $[z] \in P_u(f)$ then for some $n \geq 0$ we have $f^{km}(x) = z$ for all $k \geq n$. (Recall that if $[w] \in P_u(f)$ then $A([w],f) = \{ y \in [0,1] : f^j(y) = w \text{ for some } j \geq 0 \}$.) Hence z is not a turning point of f^{2m} and so let $[u,v]$ be the largest interval containing z on which f^{2m} is increasing; also let $\bar{u}$ and $\bar{v}$ be defined as in the proof of Lemma 5.4 but with $2m$ replacing m . We have $\bar{u} \leq z \leq \bar{v}$ and $f^{2m}([\bar{u},\bar{v}]) \subset [\bar{u},\bar{v}]$. Since $\varphi \notin f^j(J)$ for all $j \geq 0$ it follows that $f^{km}(J) \subset [\bar{u},\bar{v}]$ for all $k \geq n$, and so in particular $\bar{u} < \bar{v}$. Therefore $[\bar{u},\bar{v}]$ is a sink and of course $f^{nm}(J) \subset [\bar{u},\bar{v}]$. ◻◻

It will be convenient to divide the homtervals into two types. Let J be a homterval of $f \in S$ and let $x \in J$; we say J is a *type 1* (resp. *type 2*) homterval if $\liminf_{n\to\infty} |f^n(x)-\varphi| = 0$ (resp. $\liminf_{n\to\infty} |f^n(x)-\varphi| > 0$). *Note:* By Proposition 5.8(1) we have $\lim_{n\to\infty} |f^n(J)| = 0$, and so these definitions do not depend on the choice of $x \in J$.

Proposition 5.9 (1) If $f \in S$ satisfies (5.1) then f has no type 2 homtervals.

(2) If $f \in S_R$ has a continuous derivative in $(0,\varphi)\cup(\varphi,1)$ with $f'(z) \neq 0$ for all $z \in (0,\varphi)\cup(\varphi,1)$ and if (5.2) holds then f has no type 1 homtervals.

(3) If $f \in S_R^*$ then f has no homtervals.

Proof (3) follows directly from (1) and (2). The proof of (1) and (2) is left until Section 7. ∎∎

Now for $f \in S$ let us put

$$\mathrm{Trap}(f) = \mathrm{int}(\{ x \in [0,1] : f^m(x) \in J \text{ for some trap } J \text{ and some } m \geq 0 \}) .$$

If f is central then let

$$\mathrm{Cent}(f) = \mathrm{int}(\{ x \in [0,1] : f^m(x) \in K \text{ for some } m \geq 0 \}) ,$$

where K is the central sink of f ; if f is not central then let $\mathrm{Cent}(f) = \phi$. If $f(\varphi) > \varphi$ then put

$$\mathrm{Init}(f) = \{ x \in [0,1] : f^m(x) \in [0,\gamma) \text{ for some } m \geq 0 \} ;$$

if $f(\varphi) \leq \varphi$ then put $\mathrm{Init}(f) = \phi$. Finally, for $j = 1, 2$ let

$$\mathrm{Homt}_j(f) = \mathrm{int}(\{ x \in [0,1] : f^m(x) \in J \text{ for some type j homterval and some } m \geq 0 \}) ,$$

and put $\mathrm{Homt}(f) = \mathrm{Homt}_1(f) \cup \mathrm{Homt}_2(f)$. Note that if $f \in S_R^*$ then by Propositions 5.5 and 5.9(3) we have $\mathrm{Trap}(f) \cup \mathrm{Homt}(f) = \phi$.

Proposition 5.10 Let $f \in S$ with $f(\varphi) > \varphi$. Then

(1) $\mathrm{Trap}(f)$, $\mathrm{Cent}(f)$, $\mathrm{Init}(f)$ and $\mathrm{Homt}(f)$ are disjoint open subsets of $[0,1]$.

(2) $[\mathrm{Trap}(f) \cup \mathrm{Cent}(f) \cup \mathrm{Init}(f) \cup \mathrm{Homt}(f)] \cap \mathrm{int}(O(f))$ is dense in $\mathrm{Trap}(f) \cup \mathrm{Cent}(f) \cup \mathrm{Init}(f) \cup \mathrm{Homt}(f)$.

(3) $\mathrm{Trap}(f) \cup \mathrm{Cent}(f) \cup \mathrm{Init}(f) \subset A(f)$, but $\mathrm{Homt}(f) \cap A(f) = \phi$.

Proof (3): $\mathrm{Trap}(f) \cup \mathrm{Cent}(f) \cup \mathrm{Init}(f) \subset A(f)$ is clear, and

$\mathrm{Homt}(f)\cap A(f) = \phi$ follows from Proposition 5.8(2).

(1): Since two maximal sinks are disjoint or equal we must have $\mathrm{Trap}(f)$, $\mathrm{Cent}(f)$ and $\mathrm{Init}(f)$ are disjoint; by (3) we have that $\mathrm{Homt}(f)$ is disjoint from $\mathrm{Trap}(f)\cup\mathrm{Cent}(f)\cup\mathrm{Init}(f)$.

(2): Let

$$H(f) = \{\, x \in [0,1] : f^m(x) \in \mathrm{int}(J) \text{ for some homterval } J \text{ and some } m \geqq 0 \,\};$$

then $H(f)$ is open and $H(f) \subset \mathrm{Homt}(f)\cap O(f)$. ($H(f) \subset O(f)$ follows because $\lim_{n\to\infty}|f^n(J)| = 0$ for any homterval J.) Therefore $H(f) \subset \mathrm{Homt}(f)\cap\mathrm{int}(O(f))$. But $H(f)$ is dense in $\mathrm{Homt}(f)$ (since $\mathrm{Homt}(f)$ is open and it is easy to see that $\mathrm{Homt}(f)-H(f)$ is countable), and thus $\mathrm{Homt}(f)\cap\mathrm{int}(O(f))$ is dense in $\mathrm{Homt}(f)$. The rest follows as in the proof of Proposition 3.3. ◻◻

We can now state the main classification result for functions in $\mathcal{S}$. Note that if $f \in \mathcal{S}$ with $f(\varphi) \leq \varphi$ then f is central and $\mathrm{Cent}(f) = [0,1]$.

Theorem 5.2 Let $f \in \mathcal{S}$ with $f(\varphi) > \varphi$.

(1) If $f^2(\varphi) < \gamma$ then $\mathrm{Init}(f)\cup\mathrm{Trap}(f)\cup\mathrm{Homt}_2(f)$ is dense in $[0,1]$.

(2) If $f^2(\varphi) \geqq \gamma$ then $\mathrm{Init}(f) = [0,\gamma)\cup(\bar{\gamma},1]$ and exactly one of the following statements holds:

(5.11) f is central and $\mathrm{Trap}(f)\cup\mathrm{Cent}(f)\cup\mathrm{Homt}_2(f)$ is dense in $[\gamma,\bar{\gamma}]$.

(5.12) $\mathrm{Trap}(f)\cup\mathrm{Homt}(f)$ is dense in $[\gamma,\bar{\gamma}]$ and there exists $\varepsilon > 0$ such that $[\gamma,\bar{\gamma}] - (\mathrm{Trap}(f)\cup\mathrm{Homt}(f)) \subset \Sigma_\varepsilon(f)$.

(5.13) There exists $\varepsilon > 0$ and $\delta > 0$ such that $(\varphi-\delta,\varphi+\delta) \subset \Sigma_\varepsilon(f)$ and also $[\gamma,\bar{\gamma}] - (\mathrm{Trap}(f)\cup\mathrm{Homt}_2(f)) \subset \Sigma_\varepsilon(f)$.

(5.14) There exists a decreasing sequence $\{I^{(n)}\}_{n\geq 1}$ of closed subsets of $[\gamma,\bar{\gamma}]$, each containing φ in its interior, such that $(\{I^{(n)}\}_{n\geq 1},f)$ is an infinite register shift. Moreover, for each $n \geq 1$

$$\{ x \in [0,1] : f^k(x) \in \mathrm{int}(I^{(n)}) \text{ for some } k \geq 0 \} \cup \mathrm{Trap}(f) \cup \mathrm{Homt}_2(f)$$

is a dense open subset of $[\gamma,\bar{\gamma}]$.

Proof This is given in Section 6. ▫▫

Theorem 5.1 can be easily deduced from Theorem 5.2 once we have the following fact:

Lemma 5.5 Let $f \in S$ and $(\{I^{(n)}\}_{n\geq 1},f)$ be an infinite register shift; let $I^{(\infty)} = \bigcap_{n\geq 1} I^{(n)}$. Then $\mathrm{int}(I^{(\infty)}) \subset \mathrm{Homt}(f)$; in particular, if $\mathrm{Homt}(f) = \phi$ then $(\{I^{(n)}\}_{n\geq 1},f)$ is a proper infinite register shift.

Proof Let $J \subset I^{(\infty)}$ be a non-trivial closed interval; the intervals $\{f^n(J)\}_{n\geq 0}$ are then disjoint and so there exists $m \geq 0$ such that $\varphi \notin f^n(J)$ for all $n \geq m$. We also have $f^m(J) \cap A(f) \subset I^{(\infty)} \cap A(f) = \phi$, and hence by Proposition 5.8 $f^m(J)$ is a homterval. It immediately follows that $\mathrm{int}(I^{(\infty)}) \subset \mathrm{Homt}(f)$. ▫▫

Proof of Theorem 5.1 Let $f \in S_R^*$ with $f(z) > z$ for all $z \in (0,\varphi]$. Then $[\gamma,\bar{\gamma}] = [0,1]$ and by Propositions 5.5 and 5.9(3) we have $\mathrm{Trap}(f) \cup \mathrm{Homt}(f) = \phi$. (Thus in particular (5.12) cannot hold.) Therefore by Proposition 5.6, Theorem 5.2(2) and Lemma 5.5 we have that exactly one of (5.7), (5.8) and

(5.15) $|P_s(f)| = 1$ and $\mathrm{Cent}(f)$ is dense in $[0,1]$.

But if (5.15) holds then by Theorem 4.3 we also have (5.6), and thus exactly one of (5.6), (5.7) and (5.8) holds. □□

Remark: It is not difficult to obtain (5.6) directly from (5.15) (i.e. without using Theorem 4.3); at the end of the section we will show how this can be done.

We now examine Theorem 5.2 in more detail. First note that if $f \in \mathcal{S}$ with $f(\varphi) > \varphi$ and $f^2(\varphi) < \gamma$ then Theorem 5.2(1) and Proposition 5.10 imply that $\mathrm{int}(O(f))$ is dense in $[0,1]$; also we clearly have $(\varphi-\delta,\varphi+\delta) \subset A(f)$ for some $\delta > 0$.

Proposition 5.11 Let $f \in \mathcal{S}$ with $f(\varphi) > \varphi$ and $f^2(\varphi) \geq \gamma$.

(1) If either (5.11) or (5.12) holds then $\mathrm{int}(O(f))$ is dense in $[0,1]$.

(2) If (5.13) holds then $O(f) \cap (\varphi-\delta,\varphi+\delta) = \emptyset$ for some $\delta > 0$.

(3) If (5.14) holds and the infinite register shift $(\{I^{(n)}\}_{n\geq 1}, f)$ is proper then $O(f)$ is a residual subset of $[0,1]$.

(4) If (5.11) holds then $(\varphi-\delta,\varphi+\delta) \subset A(f)$ for some $\delta > 0$.

(5) If (5.14) holds then $\varphi \notin A(f) \cup \left(\bigcup_{\varepsilon>0} \Sigma_\varepsilon(f) \right)$.

Proof (1): Immediate from Proposition 5.10(2).

(2) and (4): These are clear.

(3): This is almost the same as the proof of Proposition 5.2.

(5): Suppose (5.14) holds and let $I^{(\infty)} = \bigcap_{n\geq 1} I^{(n)}$; we have already seen that $I^{(\infty)} \cap A(f) = \emptyset$ and so in particular $\varphi \notin A(f)$. Now let $I_0^{(n)}$ be the interval in $I^{(n)}$ containing φ . If $I_0^{(\infty)} = \bigcap_{n\geq 1} I_0^{(n)}$ is a non-trivial interval then $f((\varphi-\delta,\varphi+\delta)) \subset f(I_0^{(\infty)})$ for some $\delta > 0$ and thus $\varphi \in O(f)$ (since $\lim_{n\to\infty} |f^n(I_0^{(\infty)})| = 0$); thus in this case we have

$\varphi \notin \bigcup_{\varepsilon>0} \Sigma_\varepsilon(f)$. Assume then that $\bigcap_{n\geq 1} I_0^{(\infty)} = \{\varphi\}$ and let $\varepsilon > 0$; we can thus find $n \geq 1$ and $\delta > 0$ such that $(\varphi-\delta,\varphi+\delta) \subset I_0^{(n)} \subset (\varphi-\frac{\varepsilon}{2},\varphi+\frac{\varepsilon}{2})$ (recalling that $\varphi \in \mathrm{int}(I^{(n)})$). But $f^m(I_0^{(n)}) = I_0^{(n)}$ for some $m \geq 1$ and therefore $\{ k \geq 0 : |f^k((\varphi-\delta,\varphi+\delta))| < \varepsilon \}$ is infinite; hence $\varphi \notin \Sigma_\varepsilon(f)$. □□

Let $f \in \mathcal{S}$ with $f(\varphi) > \varphi$ and $f^2(\varphi) \geq \gamma$; we next consider the problem of how to decide which one of (5.11), (5.12), (5.13) and (5.14) holds for f . We will see that it is almost possible to determine this from just knowing what happens in any neighbourhood of φ .

First we need to modify a definition from Section 2: Fix $f \in \mathcal{S}$ and let $I_0 = [0,\varphi]$, $I_1 = [\varphi,1]$; also let

$$\Lambda = \{ \{\sigma_m\}_{m\geq 0} : \sigma_m \in \{0,1\} \text{ for each } m \geq 0 \} .$$

We defined $\sigma = \{\sigma_m\}_{m\geq 0} \in \Lambda$ to be a *code* for $x \in [0,1]$ if $f^m(x) \in I_{\sigma_m}$ for all $m \geq 0$; we also defined φ to be *regular* if φ is not periodic and if one of the two possible codes for φ is periodic. (If φ is not periodic then $f^n(\varphi) \neq \varphi$ for all $n \geq 1$ and so φ clearly has exactly two codes and these codes agree except in the zeroth component.) Let us call φ *semi-regular* if there exists a periodic $\sigma \in \Lambda$ and $\delta > 0$ such that σ is the unique code either for all the points in $(\varphi-\delta,\varphi)$ or for all the points in $(\varphi,\varphi+\delta)$. It is easy to see that if φ is either regular or a stable periodic point then φ is semi-regular.

Lemma 5.6 If $f \in \mathcal{S}$ then φ is semi-regular if and only if f is central.

Proof Suppose that f is central and let K be the central sink of f .

Then $\varphi \notin f^n(\mathrm{int}(K))$ for all $n \geq 0$ and so there exists $\sigma \in \Lambda$ such that σ is the unique code for all the points in $\mathrm{int}(K)$. But σ must be periodic because $f^m(K) \subset K$ for some $m \geq 1$, and hence φ is semi-regular. Conversely, suppose that φ is semi-regular; without loss of generality we can assume there exists $\delta > 0$ and a periodic $\sigma \in \Lambda$ such that σ is the unique code for all the points in $(\varphi-\delta,\varphi)$. Let $J = \{ x \in [0,1] : \sigma$ is a code for $x \}$; then J is a closed interval with $(\varphi-\delta,\varphi) \subset J$; it thus follows that φ is the right-hand end-point of J . Since σ is periodic it is easily checked that J is a sink, and therefore f is central. ▫▫

Proposition 5.12 Let $f \in S$ with $f(\varphi) > \varphi$ and $f^2(\varphi) \geq \gamma$. Then:

(1) (5.11) holds if and only if φ is semi-regular.

(2) (5.13) holds if and only if there exist $\varepsilon > 0$ and $\delta > 0$ such that $(\varphi-\delta,\varphi+\delta) \subset \Sigma_\varepsilon(f)$.

If in addition we know that $\varphi \notin \mathrm{Homt}_2(f)$ then:

(3) (5.14) holds if and only if $\varphi \notin A(f) \cup \left(\bigcup_{\varepsilon>0} \Sigma_\varepsilon(f) \right)$.

(4) (5.12) holds if and only if $\varphi \in A(f) \cup \left(\bigcup_{\varepsilon>0} \Sigma_\varepsilon(f) \right)$, φ is not semi-regular, and $\mathrm{int}(O(f)) \cap (\varphi-\delta,\varphi+\delta)$ is dense in $(\varphi-\delta,\varphi+\delta)$ for some $\delta > 0$.

Proof (1): Lemma 5.6 tells us that φ is semi-regular if and only if f is central. Thus by Theorem 5.2(2) it is enough to prove that if one of (5.12), (5.13) and (5.14) holds then f is not central. If f is central then by Proposition 5.11 we have $\varphi \in A(f)$ and $\mathrm{int}(O(f))$ is dense in $[0,1]$. On the other hand, if (5.14) holds then $\varphi \notin A(f)$, and if (5.13) holds then $(\varphi-\eta,\varphi+\eta) \cap O(f) = \phi$ for some $\eta > 0$. Finally, if

(5.12) holds then f is clearly not central.

(2): If (5.13) holds then of course $(\varphi-\delta,\varphi+\delta) \subset \Sigma_\varepsilon(f)$ for some $\varepsilon > 0$ and $\delta > 0$. Conversely, suppose there exist $\varepsilon > 0$ and $\delta > 0$ such that that $(\varphi-\delta,\varphi+\delta) \subset \Sigma_\varepsilon(f)$. Then as in (1) f cannot be central, and so (5.11) does not hold. However, parts (1) and (5) of Proposition 5.11 show that (5.12) and (5.14) also cannot hold, and thus by Theorem 5.2 we must have (5.13)

Now let us suppose that $\varphi \notin \mathrm{Homt}_2(f)$; we thus in fact have $\varphi \notin \mathrm{Homt}(f)$ because $\varphi \in \mathrm{Homt}_1(f)$ is never possible.

(3): If (5.14) holds then by Proposition 5.11(5) we have $\varphi \notin A(f) \cup \left(\bigcup_{\varepsilon>0} \Sigma_\varepsilon(f)\right)$. Conversely, suppose that $\varphi \notin A(f) \cup \left(\bigcup_{\varepsilon>0} \Sigma_\varepsilon(f)\right)$; then clearly neither (5.11) nor (5.13) holds. If (5.12) held then, since $\varphi \notin \mathrm{Homt}(f)$ and $\mathrm{Trap}(f) \subset A(f)$, we would have

$$\varphi \in \mathrm{Trap}(f) \cup \left(\bigcup_{\varepsilon>0} \Sigma_\varepsilon(f)\right) \subset A(f) \cup \left(\bigcup_{\varepsilon>0} \Sigma_\varepsilon(f)\right) ;$$

thus (5.12) does not hold and hence by Theorem 5.2 we must have (5.14).

(4): This is now clear. ☐☐

Note that we automatically have $\varphi \notin \mathrm{Homt}_2(f)$ when f satisfies (5.1), since in this case $\mathrm{Homt}_2(f) = \phi$. Thus if we have $f \in S$ with $f(\varphi) > \varphi$ and $f^2(\varphi) \geqq \gamma$ and if also f satisfies (5.1) then we can determine which of (5.11), (5.12), (5.13) and (5.14) holds from just knowing what happens in any neighbourhood of φ .

We end this section by showing how it is possible in the proof of Theorem 5.1 to obtain (5.6) directly from (5.15) (i.e. without making use of Theorem 4.3). Thus let $f \in S_R^*$ satisfy (5.15) and let K be the central sink of f ; without loss of generality we can assume φ is the right-hand end-point of K . Let m be the smallest positive integer such that both $f^m(K) \subset K$ and f^m is increasing on K ; let $[u,\varphi]$ be the largest interval containing K on which f^m is increasing. Then we have $K = [\overline{u},\varphi]$, where

$$\overline{u} = \begin{cases} \text{the smallest fixed point of } f^m \text{ in } [u,\varphi] & \text{if } f^m(u) < u , \\ u & \text{if } f^m(u) \geq u . \end{cases}$$

If $u < \overline{u}$ then (since f has property R) Proposition 4.6 implies that there exists $w \in [\overline{u},\varphi]$ with $f^m(w) = w$ such that $f^m(z) > z$ for all $z \in (\overline{u},w)$ and $f^m(z) < z$ for all $z \in (w,\varphi]$.

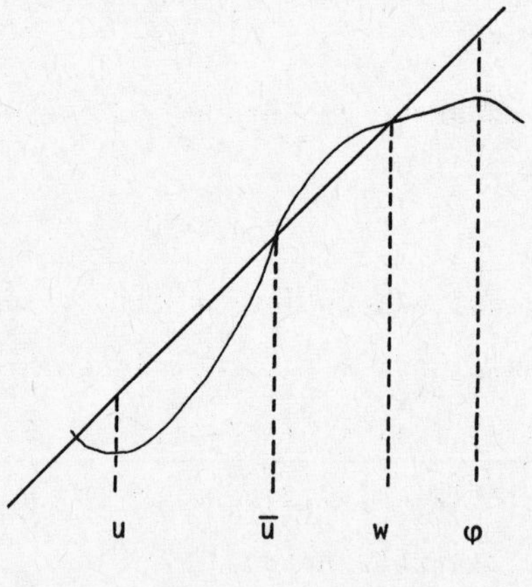

It follows that $[w] \in P_s(f)$ (and so $P_s(f) = \{[w]\}$) and also that $C(f) \subset A([w],f)$, where

$$C(f) = \{ x \in [0,1] : f^k(x) \in (\bar{u},\varphi) \text{ for some } k \geq 0 \} .$$

But Cent(f)-C(f) is countable, and so by (5.15) C(f) is a dense open subset of [0,1] ; therefore in this case (5.6) holds. If $u = \bar{u}$ then, since $u \neq 0$ and $f^m([u,\varphi]) \subset [u,\varphi]$, we must have $f^j(u) = \varphi$ with $2j = m$. There thus exists $w \in (u,\varphi)$ with $f^j(w) = w$ and Proposition 4.6 now implies that either

(i) $f^m(z) > z$ for all $z \in [u,w)$ and $f^m(z) < z$ for all $z \in (w,\varphi]$,

or

(ii) there exist $s \in [u,w)$ and $t \in (w,\varphi]$ with $f^j(s) = t$ and $f^j(t) = s$ such that $f^m(z) > z$ for all $z \in [u,s)\cup(w,t)$ and $f^m(z) < z$ for all $z \in (s,w)\cup(t,\varphi]$.

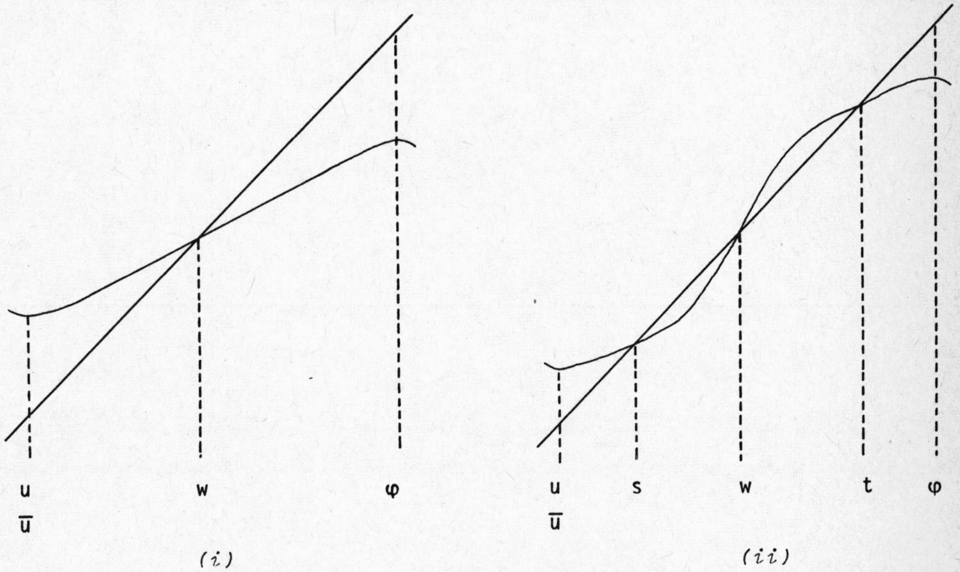

In case *(i)* we have $[w] \in P_s(f)$ and $Cent(f) = A([w],f)$, and so (5.6) holds. In case *(ii)* we have $[s] \in P_s(f)$ and $C'(f) \subset A([s],f)$, where

$$C'(f) = \{ x \in [0,1] : f^k(x) \in (u,w)\cup(w,\varphi) \text{ for some } k \geq 0 \} .$$

But Cent(f)-C'(f) is countable and hence by (5.15) C'(f) is a dense open subset of [0,1] ; thus again (5.6) holds.

Theorem 5.1 is a combination of results due to Guckenheimer and Misiurewicz (Guckenheimer (1979), Misiurewicz (1980)). The definition of sensitive dependence which we have used is stronger than that introduced by Guckenheimer: he defines f to have sensitive dependence to initial conditions if $\lambda(\Sigma_\varepsilon(f)) > 0$ for some $\varepsilon > 0$. The actual results of Guckenheimer and Misiurewicz involve statements about Lebesgue measure and are given in Section 9. The formulation of (5.8) is based on Theorem 2.6 in Collet, Eckmann and Lanford (1980).

6. REDUCTIONS

This section is taken up with the proof of Theorem 5.2. The proof is based on a construction which involves what we call a reduction. Let $f \in \mathcal{S}$, $0 \leq c < \varphi < d \leq 1$ and $m > 1$; we call $\Gamma = ([c,d],f^m)$ a *reduction* of $([0,1],f)$ if

(6.1) $f^m([c,d]) \subset [c,d]$,

(6.2) the intervals $(c,d), f((c,d)),\ldots,f^{m-1}((c,d))$ are disjoint.

We call m the *order* of Γ .

Lemma 6.1 Let $f \in \mathcal{S}$ and $([c,d],f^m)$ be a reduction of $([0,1],f)$. Then φ is the only turning point of f^m in $[c,d]$, and so in particular f^m is strictly monotone on each of the intervals $[c,\varphi]$ and $[\varphi,d]$.

Proof Let $z \in [c,d]$ be a turning point of f^m ; then by Proposition 2.1 there exists $0 \leq k < m$ with $f^k(z) = \varphi$. If $0 < k < m$ then by (6.2) we have $f^k((c,d)) \cap (c,d) = \phi$ and thus also $f^k([c,d]) \cap (c,d) = \phi$. Hence $k = 0$ and therefore $z = \varphi$. ⊞

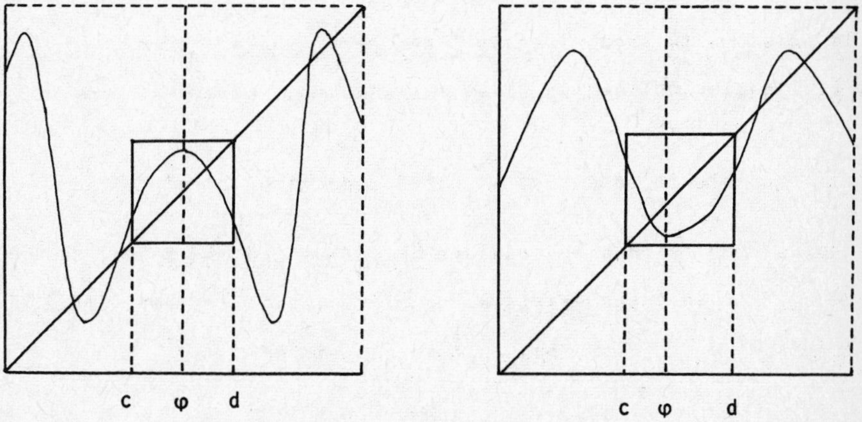

Let $f \in \mathcal{S}$ and $\Gamma = ([c,d],f^m)$ be a reduction of $([0,1],f)$; let g denote the restriction of f^m to $[c,d]$. By Lemma 6.1 g is either a scaled down or a scaled down and turned upside down version of a function from $\mathcal{S}$. We can thus transform g into an element of $\mathcal{S}$ by a linear change of variables. Let us denote the element of $\mathcal{S}$ obtained in this way by $\mathcal{U}_\Gamma f$, so $\mathcal{U}_\Gamma f = \Psi_\Gamma \circ g \circ \Psi_\Gamma^{-1}$, where $\Psi_\Gamma : [c,d] \to [0,1]$ is given by

$$\Psi_\Gamma(x) = \begin{cases} \dfrac{x-c}{d-c} & \text{if } f^m \text{ is increasing on } [c,\varphi] , \\ \dfrac{d-x}{d-c} & \text{if } f^m \text{ is increasing on } [\varphi,d] . \end{cases}$$

Thus in fact we have

$$(\mathcal{U}_\Gamma f)(x) = \begin{cases} \dfrac{1}{d-c}\Big(f^m((d-c)x+c) - c \Big) & \text{if } f^m \text{ is increasing on } [c,\varphi] , \\ \dfrac{1}{d-c}\Big(d - f^m(d-(d-c)x) \Big) & \text{if } f^m \text{ is increasing on } [\varphi,d] . \end{cases}$$

The idea behind the proof of Theorem 5.2 is roughly the following: For each $f \in \mathcal{S}$ we will see that either

(i) it is possible to show directly that one of the statements in Theorem 5.2, excluding (5.14), holds,

or

(ii) it is possible to construct a "nice" reduction Γ of $([0,1],f)$.

If *(ii)* holds for f then we consider $\mathcal{U}_\Gamma f$ instead of f ; if *(ii)* holds for $\mathcal{U}_\Gamma f$ then there exists a "nice" reduction Γ' of $([0,1],\mathcal{U}_\Gamma f)$ and we consider $\mathcal{U}_{\Gamma'}(\mathcal{U}_\Gamma f)$. Thus in this way either *(iii)* after finitely many iterations we end up with a function $g = U_{\Gamma_n}(\ldots U_{\Gamma_2}(U_{\Gamma_1} f)\ldots) \in \mathcal{S}$ for which *(i)* holds,

or

(iv) this reduction procedure can be repeated indefinitely.

If *(iii)* holds then we will show that f satisfies the condition in Theorem 5.2 which is satisfied by g. This leaves us with the functions for which *(iv)* holds, and we will show that these functions satisfy (5.14).

We now start the detailed implementation of the above program. Let $f \in \mathcal{S}$ and $\Gamma = ([c,d],f^m)$ be a reduction of $([0,1],f)$; put

$$e_\Gamma = \begin{cases} c & \text{if } f^m \text{ is increasing on } [c,\varphi] , \\ d & \text{if } f^m \text{ is increasing on } [\varphi,d] . \end{cases}$$

Lemma 6.2 $e_\Gamma \in (0,1)$.

Proof It is easy to see that if $f(\varphi) \leq \varphi$ then no reduction can exist; thus $f(\varphi) > \varphi$ and so f has a unique fixed point $\beta \in (\varphi,1)$. Clearly $d \leq \beta$ and hence in particular $e_\Gamma \neq 1$. Now for any $f \in \mathcal{S}$ with $f(\varphi) > \varphi$ we have $f^n(\varphi) \geq f^2(\varphi)$ for all $n \geq 2$; thus if $e_\Gamma = 0$ then $m = 2$, and this is not possible (since if $m = 2$ then $e_\Gamma = d$). ☐☐

If $e_\Gamma = c$ (resp. $e_\Gamma = d$) then let $[u,\varphi]$ (resp. $[\varphi,v]$) be the largest interval containing e_Γ on which f^m is increasing; by Lemmas 6.1 and 6.2 we have $u < e_\Gamma$ (resp. $e_\Gamma < v$). We call the reduction Γ *simple* if $f^m(c) = f^m(d) = e_\Gamma$ and if $f^m(z) < z$ for all $z \in [u,e_\Gamma)$ (resp. $f^m(z) > z$ for all $z \in (e_\Gamma,v]$). (See the diagrams at the top of the next page.)

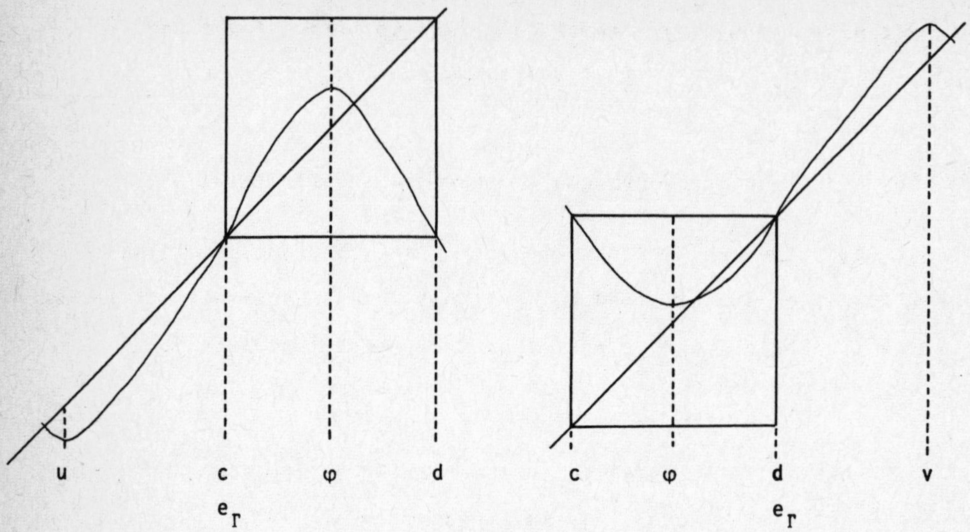

Remark: If $f \in S$ and $m > 1$ then there exists at most one simple reduction of $([0,1],f)$ having order m.

Let $f \in S$ and $\Gamma = ([c,d],f^m)$ be a reduction of $([0,1],f)$. Put

$$C_\Gamma = \{ x \in [0,1] : f^n(x) \in (c,d) \text{ for some } n \geq 0 \},$$

and let $B_\Gamma = \text{int}([0,1]-C_\Gamma)$; thus B_Γ and C_Γ are both open and $B_\Gamma \cup C_\Gamma$ is a dense open subset of $[0,1]$.

We can now give (in Propositions 6.1, 6.2 and 6.3) the precise formulation of *(i)* and *(ii)* above. If $f \in S$ with $f(\varphi) > \varphi$ then let β denote the unique fixed point of f in $(\varphi,1)$; if $f(0) < \beta$ then let α be the unique point in $(0,\varphi)$ with $f(\alpha) = \beta$, if $f(0) \geq \beta$ then let $\alpha = 0$.

Proposition 6.1 Let $f \in S$ with $f(\varphi) > \varphi$ and $f^2(\varphi) \geq \alpha$. Then:

(1) $\Gamma = ([\alpha,\beta],f^2)$ is a reduction of $([0,1],f)$ with $B_\Gamma = [0,\gamma) \cup (\overline{\gamma},1]$

and $[0,1]-(B_\Gamma \cup C_\Gamma) \subset N \cup \{ z \in [0,1] : f^n(z) = \beta \text{ for some } n \geq 0 \}$,

where $N = \begin{cases} \{\gamma,\bar{\gamma}\} & \text{if } f(\gamma) = \gamma \text{ and } f(\bar{\gamma}) = \gamma , \\ \{\gamma\} & \text{if } f(\gamma) = \gamma \text{ and } f(\bar{\gamma}) \neq \gamma , \\ \phi & \text{otherwise.} \end{cases}$

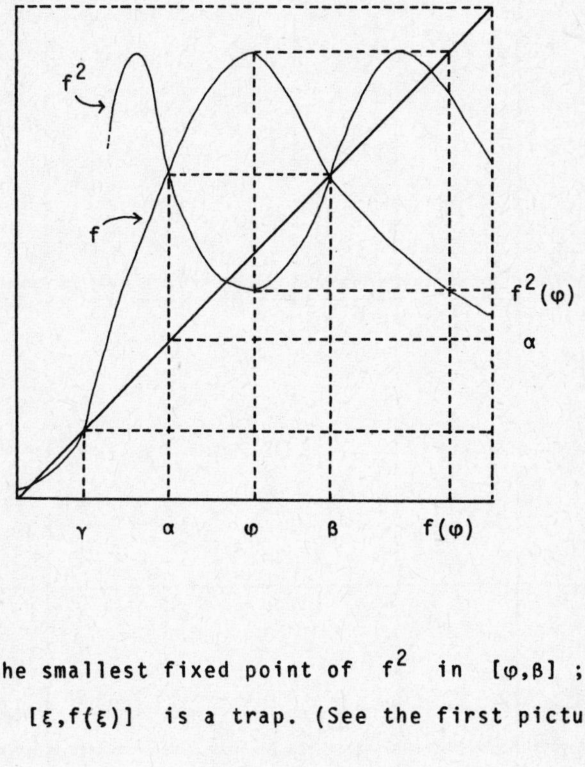

(2) Let ξ be the smallest fixed point of f^2 in $[\varphi,\beta]$; if $f^2(\varphi) < \varphi$ and $\xi < \beta$ then $[\xi,f(\xi)]$ is a trap. (See the first picture on the next page.)

Proposition 6.2 Let $f \in S$ with $f(\varphi) > \varphi$ and $f^2(\varphi) < \gamma$. Then $\mathrm{Init}(f) \cup \mathrm{Trap}(f) \cup \mathrm{Homt}_2(f)$ is dense in $[0,1]$ and

$[0,1] - (\mathrm{Init}(f) \cup \mathrm{Trap}(f) \cup \mathrm{Homt}_2(f)) \subset \Sigma_\varepsilon(f)$, where $\varepsilon = \min\{\beta-\alpha, f(\varphi)-\beta\}$.

(See the second picture on the next page.)

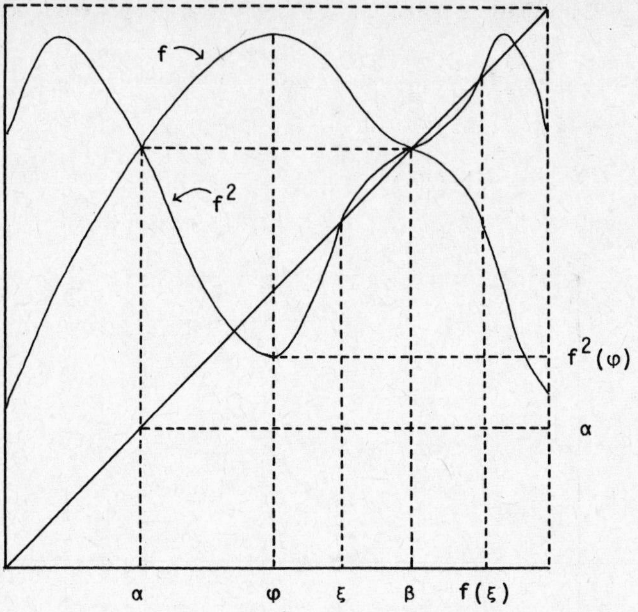
f
f²
f²(φ)
α
α
φ
ξ
β
f(ξ)

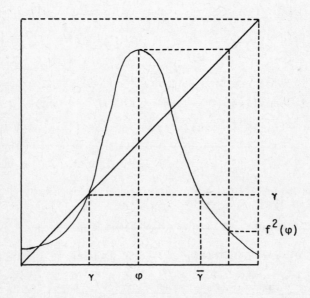

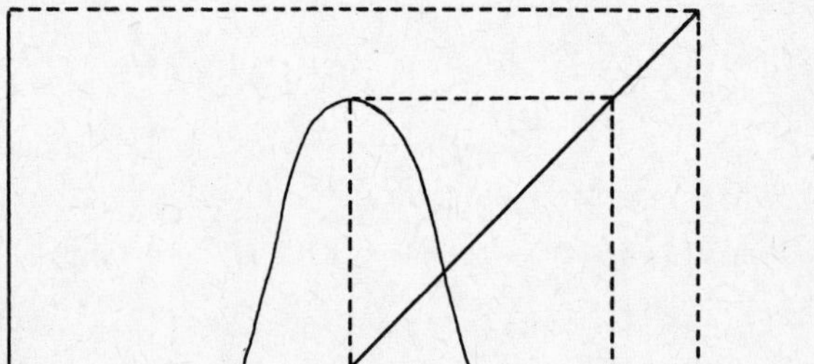
γ
f²(φ)
γ
φ
γ̄

Proposition 6.3 Let $f \in S$ with $f(\varphi) > \varphi$ and $\gamma \leq f^2(\varphi) < \alpha$. Then $\text{Init}(f) = [0,\gamma)\cup(\bar{\gamma},1]$ and one of the following holds:

(6.3) $\text{Trap}(f)\cup\text{Homt}(f)$ is dense in $[\gamma,\bar{\gamma}]$ and $[\gamma,\bar{\gamma}] - (\text{Trap}(f)\cup\text{Homt}(f)) \subset \Sigma_\varepsilon(f)$, where $\varepsilon = \min\{\beta-\alpha, f(\varphi)-\beta\}$.

(6.4) $\text{Trap}(f)\cup\text{Homt}(f) = \phi$ and $[\gamma,\bar{\gamma}] \subset \Sigma_\varepsilon(f)$ (with ε as in (6.3)).

(6.5) There exists a simple reduction $\Gamma = ([c,d],f^m)$ of $([0,1],f)$ with $B_\Gamma \subset \text{Init}(f)\cup\text{Trap}(f)\cup\text{Homt}_2(f)$ and $[0,1] - (B_\Gamma \cup C_\Gamma) \subset \Sigma_\varepsilon(f)$ (with ε again as in (6.3)).

(See the pictures on the next page. The second picture shows the simplest example of (6.5) holding; this occurs with $m = 3$.)

Before proving Propositions 6.1, 6.2 and 6.3 we will show how they can be used to obtain Theorem 5.2. First note that Theorem 5.2(1) is part of Proposition 6.2 and so we only have to worry about Theorem 5.2(2); furthermore, if $f \in S$ with $f(\varphi) > \varphi$ and $f^2(\varphi) \geq \gamma$ then it is clear that $\text{Init}(f) = [0,\gamma)\cup(\bar{\gamma},1]$ (because $f([\gamma,\bar{\gamma}]) \subset [\gamma,\bar{\gamma}]$).

Lemma 6.3 Let $f \in S$ with $f(\varphi) > \varphi$ and $f^2(\varphi) \geq \gamma$. Then at most one of (5.11), (5.12), (5.13) and (5.14) holds.

Proof We have already seen in the proof of Proposition 5.12(1) that if one of (5.12), (5.13) and (5.14) holds then f is not central. Also parts (1) and (5) of Proposition 5.11 show that if one of (5.11), (5.12) and (5.14) holds then (5.13) does not hold. Thus we need only prove that (5.12) and (5.14) cannot both hold. For each $\varepsilon > 0$ let

$$W_\varepsilon = [\gamma,\bar{\gamma}] - (\text{Trap}(f)\cup\text{Homt}(f)\cup\Sigma_\varepsilon(f)) ;$$

if (5.12) holds then of course $W_\varepsilon = \phi$ for some $\varepsilon > 0$. We will show

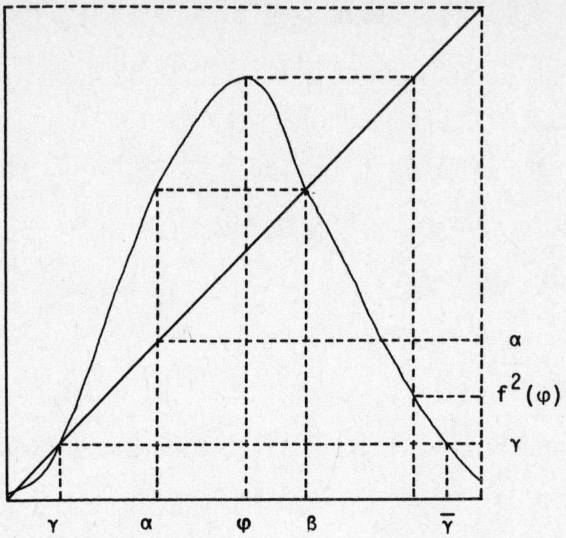
α
$f^2(\varphi)$
γ
γ
α
φ
β
$\bar{\gamma}$

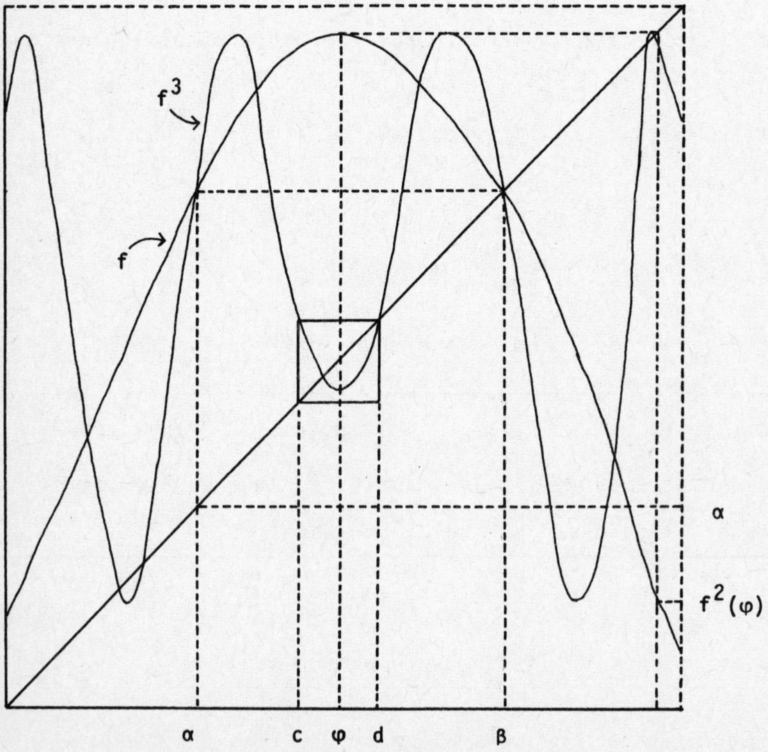
f^3
f
α
$f^2(\varphi)$
α
c
φ
d
β

that if (5.14) holds then $\bigcap_{\varepsilon>0} W_\varepsilon$ is uncountable, and this gives us the required difference between (5.12) and (5.14). Thus suppose (5.14) holds and put $I^{(\infty)} = \bigcap_{n\geq 1} I^{(n)}$; for $x \in I^{(\infty)}$ and $n \geq 1$ let $I_x^{(n)}$ be the interval in $I^{(n)}$ containing x . Also let

$I_*^{(\infty)} = \{ x \in I^{(\infty)} : \bigcap_{n\geq 1} I_x^{(n)} = \{x\} \}$ and

$I_{**}^{(\infty)} = \{ x \in I_*^{(\infty)} : x \in \mathrm{int}(I_x^{(n)})$ for all $n \geq 1 \}$; it is easy to see that $I_{**}^{(\infty)}$ is an uncountable subset of $I^{(\infty)}$. If $x \in I_{**}^{(\infty)}$ then $f^k(x) \in I_*^{(\infty)}$ for all $k \geq 0$; if $y \in I_*^{(\infty)}$ and J is a non-trivial interval with $y \in \mathrm{int}(J)$ then clearly the intervals $\{f^n(J)\}_{n\geq 0}$ are not disjoint, and so in particular Proposition 5.8(1) implies that J is not a homterval. Thus $I_{**}^{(\infty)} \cap H(f) = \phi$, where

$$H(f) = \{ x \in [0,1] : f^m(x) \in \mathrm{int}(J) \text{ for some homterval } J \text{ and some } m \geq 0 \} .$$

But $\mathrm{Homt}(f) - H(f)$ is countable (cf. the proof of Proposition 5.10), and hence $I_{**}^{(\infty)} \cap \mathrm{Homt}(f)$ is countable. Now the proof of Proposition 5.11(5) shows that $I_{**}^{(\infty)} \cap \left(\bigcup_{\varepsilon>0} \Sigma_\varepsilon(f) \right) = \phi$, and we clearly have $I_{**}^{(\infty)} \cap \mathrm{Trap}(f) = \phi$ (since $\mathrm{Trap}(f) \subset A(f)$ and $A(f) \cap I^{(\infty)} = \phi$). Therefore $I_{**}^{(\infty)} - \bigcap_{\varepsilon>0} W_\varepsilon$ is countable, and this implies that $\bigcap_{\varepsilon>0} W_\varepsilon$ is uncountable. ◻◻

By Lemma 6.3 and the remarks preceding it the proof of Theorem 5.2 is reduced to showing that if $f \in \mathcal{S}$ with $f(\varphi) > \varphi$ and $f^2(\varphi) \geq \gamma$ then one of (5.11), (5.12), (5.13) and (5.14) holds.

We now use Proposition 6.3 to partition $\mathcal{S}$ into six parts: we can write $\mathcal{S}$ as a disjoint union $\bigcup_{k=1}^{6} \mathcal{S}_k$, where

$\mathcal{S}_1 = \{ f \in \mathcal{S} : f(\varphi) \leq \varphi \}$, $\mathcal{S}_2 = \{ f \in \mathcal{S} : f(\varphi) > \varphi$ and $f^2(\varphi) \geq \alpha \}$,

$\mathcal{S}_3 = \{ f \in \mathcal{S} : f(\varphi) > \varphi , \gamma \leq f^2(\varphi) < \alpha$ and (6.5) holds $\}$,

$\mathcal{S}_4 = \{ f \in \mathcal{S}-\mathcal{S}_3 : f(\varphi) > \varphi , \gamma \leq f^2(\varphi) < \alpha$ and (6.3) holds $\}$,

$\mathcal{S}_5 = \{ f \in \mathcal{S}-\mathcal{S}_3 : f(\varphi) > \varphi , \gamma \leq f^2(\varphi) < \alpha$ and (6.4) holds $\}$,

$\mathcal{S}_6 = \{ f \in \mathcal{S} : f(\varphi) > \varphi$ and $f^2(\varphi) < \gamma \}$.

For $f \in \mathcal{S}_2$ we define $\mathcal{T}f : [0,1] \to [0,1]$ by

$$(\mathcal{T}f)(x) = \frac{1}{\beta-\alpha} \left(\beta - f^2(\beta-(\beta-\alpha)x) \right),$$

i.e. $\mathcal{T}f = \mathcal{U}_\Gamma f$, where $\Gamma = ([\alpha,\beta],f^2)$ is the reduction of $([0,1],f)$ given by Proposition 6.1(1). This gives us a mapping $\mathcal{T} : \mathcal{S}_2 \to \mathcal{S}$. Similarly, for $f \in \mathcal{S}_3$ we let $\mathcal{U}f = \mathcal{U}_\Gamma f$, where Γ is the simple reduction of smallest order satisfying (6.5); this gives us a mapping $\mathcal{U} : \mathcal{S}_3 \to \mathcal{S}$. Now define $\mathcal{V} : \mathcal{S} \to \mathcal{S}$ by

$$\mathcal{V}f = \begin{cases} f & \text{if } f \in \mathcal{S}_1 \cup \mathcal{S}_4 \cup \mathcal{S}_5 \cup \mathcal{S}_6 , \\ \mathcal{T}f & \text{if } f \in \mathcal{S}_2 , \\ \mathcal{U}f & \text{if } f \in \mathcal{S}_3 , \end{cases}$$

and for $n \geq 0$ let $\mathcal{V}^n : \mathcal{S} \to \mathcal{S}$ be given inductively by $\mathcal{V}^0 f = f$, $\mathcal{V}^1 f = \mathcal{V}f$ and $\mathcal{V}^n f = \mathcal{V}(\mathcal{V}^{n-1} f)$.

Theorem 5.2(2) is now a direct consequence of Lemma 6.3 and the next result.

Proposition 6.4 Let $f \in S$ with $f(\varphi) > \varphi$ and $f^2(\varphi) \geqq \gamma$ (i.e. $f \in S_2 \cup S_3 \cup S_4 \cup S_5$). Then we have

(1) If $V^n f \in S_1$ for some $n \geqq 0$ then (5.11) holds.

(2) If $V^n f \in S_4$ for some $n \geqq 0$ then (5.12) holds.

(3) If $V^n f \in S_5$ for some $n \geqq 0$ then (5.13) holds.

(4) If $V^n f \in S_6$ for some $n \geqq 0$ then (5.12) holds.

(5) If $V^n f \in S_2 \cup S_3$ for all $n \geqq 0$ then (5.14) holds.

Proof For $n \geqq 1$ and $k = 1, 4, 5$ and 6 let

$$S_k^{(n)} = \{ f \in S : V^n f \in S_k \text{ but } V^{n-1} f \notin S_k \} ,$$

also put $S_k^{(0)} = S_k$. Thus

$$\bigcup_{n \geqq 0} S_k^{(n)} = \{ f \in S : V^n f \in S_k \text{ for some } n \geqq 0 \} ,$$

and if $f \in S_k^{(n)}$ then $V^j f \in S_2 \cup S_3$ for $j = 0,\dots,n-1$. Now consider $f \in S_k^{(n)}$ with $n \geqq 1$ and $k = 1, 4, 5$ or 6 ; then we have $f \in S_2 \cup S_3$ and $Vf \in S_k^{(n-1)}$. Therefore (1), (2), (3) and (4) of Proposition 6.4 follow immediately by induction from the next result (noting that if $f \in S_4$ (resp. $f \in S_5$) then (5.12) (resp. (5.13)) holds).

Proposition 6.5 Let $f \in S_2 \cup S_3$ and put $g = Vf$. Then we have:

(1) If $g \in S_1$ then f satisfies (5.11); furthermore, if $g \notin S_1$ and g satisfies (5.11) then again f satisfies (5.11).

(2) If $g \notin S_1$ and g satisfies (5.12) then f satisfies (5.12).

(3) If $g \notin \mathcal{S}_1$ and g satisfies (5.13) then f satisfies (5.13).

(4) If $g \in \mathcal{S}_6$ then f satisfies (5.12).

Proof (of Proposition 6.5) First we need the following:

Lemma 6.4 Let $f \in \mathcal{S}_2$ (resp. $f \in \mathcal{S}_3$) and let $\Gamma = ([c,d],f^m)$ be the reduction of $([0,1],f)$ given by Proposition 6.1 (resp. Proposition 6.3); put $g = \mathcal{U}_\Gamma f$ (and so $g = \mathcal{V}f$). Then we have:

(1) If g is central then f is central and

$$\Psi_\Gamma^{-1}(\mathrm{Cent}(g)) \cap (c,d) \subset \mathrm{Cent}(f) .$$

(2) $\Psi_\Gamma^{-1}(\mathrm{Init}(g) \cup \mathrm{Trap}(g)) \cap (c,d) \subset \mathrm{Trap}(f)$.

(3) $\Psi_\Gamma^{-1}(\mathrm{Homt}_j(g)) \cap (c,d) \subset \mathrm{Homt}_j(f)$ for $j = 1$ and 2 .

(4) For each $\varepsilon > 0$ there exists $\eta > 0$ so that $\Psi_\Gamma^{-1}(\Sigma_\varepsilon(g)) \subset \Sigma_\eta(f)$.

Proof (of Lemma 6.4) (1): If J is a sink of g then clearly $\Psi_\Gamma^{-1}(J)$ is a sink of f ; furthermore, if $\varphi(g)$ is an end-point of J then $\varphi(f)$ is an end-point of $\Psi_\Gamma^{-1}(J)$. Thus if g is central with central sink K' then f is central, and if K is the central sink of f then $\Psi_\Gamma^{-1}(K') \subset K$. Therefore

$$\begin{aligned}
&\Psi_\Gamma^{-1}(\{ y \in [0,1] : g^n(y) \in K' \text{ for some } n \geq 0 \}) \\
&\subset \{ x \in [0,1] : f^{mn}(x) \in \Psi_\Gamma^{-1}(K') \text{ for some } n \geq 0 \} \\
&\subset \{ x \in [0,1] : f^j(x) \in K \text{ for some } j \geq 0 \} ,
\end{aligned}$$

and hence $\Psi_\Gamma^{-1}(\mathrm{Cent}(g)) \cap (c,d) \subset \mathrm{Cent}(f)$ (since for any $A \subset [0,1]$ we have $\Psi_\Gamma^{-1}(\mathrm{int}(A)) \cap (c,d) \subset \mathrm{int}(\Psi_\Gamma^{-1}(A))$).

(2): If J is a trap of g then it is easy to see that $\Psi_\Gamma^{-1}(J)$ is a trap of f ; as in (1) this gives us $\Psi_\Gamma^{-1}(\mathrm{Trap}(g))\cap(c,d) \subset \mathrm{Trap}(f)$. Now suppose $g \notin S_1$ with $\gamma(g) > 0$; then Proposition 6.1(2) (in the case $f \in S_2$) or the fact that Γ is simple (in the case $f \in S_3$) implies that there exists a trap J (of f) with $\Psi_\Gamma^{-1}([0,\gamma(g)]) \subset J$. Thus $\Psi_\Gamma^{-1}(\mathrm{Init}(g)) \subset \{ x \in [0,1] : f^k(x) \in J$ for some $k \geq 0 \}$; but $\Psi_\Gamma^{-1}(\mathrm{Init}(g))\cap(c,d)$ is an open subset of $[0,1]$ and hence $\Psi_\Gamma^{-1}(\mathrm{Init}(g))\cap(c,d) \subset \mathrm{Trap}(f)$. This is also trivially true if either $g \notin S_1$ and $\gamma(g) = 0$ or if $g \in S_1$, since in these cases we have $\mathrm{Init}(g) = \phi$.

(3): This follows as in (1) because if J is a type 1 (resp. type 2) homterval of g then $\Psi_\Gamma^{-1}(J)$ is a type 1 (resp. type 2) homterval of f .

(4): Let $\xi = (d-c)\varepsilon$ and for $0 \leq k < m$ let

$$\varepsilon_k = \inf\{ |f^k([z,z+\xi])| : c \leq z \leq d \} ;$$

by compactness we have $\varepsilon_k > 0$. Put $\eta = \min_{0\leq k<m} \varepsilon_k$. Now take $x \in \Psi_\Gamma^{-1}(\Sigma_\varepsilon(g))$ and let $y = \Psi_\Gamma(x)$, thus $y \in \Sigma_\varepsilon(g)$. Let $\delta > 0$ and put $\sigma = (d-c)\delta$; there then exists $n \geq 0$ such that for all $k \geq n$ we have $|g^k((y-\sigma,y+\sigma))| \geq \varepsilon$, and thus $|f^{mk}((x-\delta,x+\delta))| \geq (d-c)\varepsilon = \xi$ for all $k \geq n$. Therefore $|f^j((x-\delta,x+\delta))| \geq \eta$ for all $j \geq nm$, and hence $x \in \Sigma_\eta(f)$. ▫▫

We now apply Lemma 6.4 to the proof of Proposition 6.5. Let $f \in S_2 \cup S_3$ and let $\Gamma = ([c,d],f^m)$ be the reduction of $([0,1],f)$ given by either Proposition 6.1 or Proposition 6.3; put $g = U_\Gamma f$. Note that in both cases we have

(6.6) $B_\Gamma \subset \mathrm{Init}(f)\cup\mathrm{Trap}(f)\cup\mathrm{Homt}_2(f)$.

(1): By Lemma 6.4(1) we have f is central whenever g is. Suppose $g \in S_1$; then $\mathrm{Cent}(g) = [0,1]$, and so again using Lemma 6.4(1) we have $(c,d) = \Psi_\Gamma^{-1}([0,1])\cap(c,d) \subset \mathrm{Cent}(f)$. Thus also $C_\Gamma \subset \mathrm{Cent}(f)$, and together with (6.6) this gives us that $\mathrm{Init}(f)\cup\mathrm{Trap}(f)\cup\mathrm{Cent}(f)\cup\mathrm{Homt}_2(f)$ is dense in $[0,1]$, i.e. $\mathrm{Trap}(f)\cup\mathrm{Cent}(f)\cup\mathrm{Homt}_2(f)$ is dense in $[\gamma,\overline{\gamma}]$. Suppose now that $g \notin S_1$ and g satisfies (5.11) ; we thus have $\mathrm{Init}(g)\cup\mathrm{Trap}(g)\cup\mathrm{Cent}(g)\cup\mathrm{Homt}_2(g)$ is dense in $[0,1]$, and hence $\Psi_\Gamma^{-1}(\mathrm{Init}(g)\cup\mathrm{Trap}(g)\cup\mathrm{Cent}(g)\cup\mathrm{Homt}_2(g))\cap(c,d)$ is dense in (c,d) . Therefore by Lemma 6.4 $(\mathrm{Trap}(f)\cup\mathrm{Cent}(f)\cup\mathrm{Homt}_2(f))\cap(c,d)$ is dense in (c,d) and it thus follows that $(\mathrm{Trap}(f)\cup\mathrm{Cent}(f)\cup\mathrm{Homt}_2(f))\cap C_\Gamma$ is dense in C_Γ . As before this implies that $\mathrm{Trap}(f)\cup\mathrm{Cent}(f)\cup\mathrm{Homt}_2(f)$ is dense in $[\gamma,\overline{\gamma}]$.

(3): Suppose $g \notin S_1$ and that g satisfies (5.13). There thus exist $\varepsilon > 0$ and $\delta > 0$ such that $(\varphi(g)-\delta,\varphi(g)+\delta) \subset \Sigma_\varepsilon(g)$ and $[\gamma(g),\overline{\gamma}(g)] - (\mathrm{Trap}(g)\cup\mathrm{Homt}_2(g)) \subset \Sigma_\varepsilon(g)$. Let us put

$$Y(g) = [0,1] - (\mathrm{Init}(g)\cup\mathrm{Trap}(g)\cup\mathrm{Homt}_2(g)) ,$$

so $Y(g) \subset \Sigma_\varepsilon(g)$. Also let $\eta > 0$ be given by Lemma 6.4(4); then by Lemma 6.4 we have

$$\begin{aligned}(c,d) &= \Psi_\Gamma^{-1}(\mathrm{Init}(g)\cup\mathrm{Trap}(g)\cup\mathrm{Homt}_2(g)\cup Y(g))\cap(c,d)\\ &\subset \mathrm{Trap}(f)\cup\mathrm{Homt}_2(f)\cup\Psi_\Gamma^{-1}(Y(g)) \subset \mathrm{Trap}(f)\cup\mathrm{Homt}_2(f)\cup\Sigma_\eta(f) ,\end{aligned}$$

and therefore $C_\Gamma \subset \mathrm{Trap}(f)\cup\mathrm{Homt}_2(f)\cup\Sigma_\eta(f)$. Furthermore, if $\sigma = (d-c)\delta$

then $(\varphi-\sigma,\varphi+\sigma) \subset \Sigma_\eta(f)$. Now if $f \in S_3$ then there exists $\nu > 0$ such that $[0,1] - (B_\Gamma \cup C_\Gamma) \subset \Sigma_\nu(f)$, and so in this case (5.13) holds with $\varepsilon = \min\{\nu,\eta\}$ and $\delta = \sigma$. If $f \in S_2$ then we have

$$[\gamma,\bar{\gamma}] - (\mathrm{Trap}(f) \cup \mathrm{Homt}_2(f)) \subset \Sigma_\eta(f) \cup N \cup Z ,$$

where $N = \begin{cases} \{\gamma,\bar{\gamma}\} & \text{if } f(\gamma) = \gamma \text{ and } f(\bar{\gamma}) = \gamma , \\ \{\gamma\} & \text{if } f(\gamma) = \gamma \text{ and } f(\bar{\gamma}) \neq \gamma , \\ \phi & \text{otherwise,} \end{cases}$ and where

$Z = \{ x \in [0,1] : f^n(x) = \beta \text{ for some } n \geq 0 \}$. If $\gamma \in N$ then clearly $\gamma \in \Sigma_{\varphi-\gamma}(f)$, and thus in all cases we have $N \subset \Sigma_{\varphi-\gamma}(f)$. But Proposition 6.1(2) gives us that either $\beta \in \mathrm{Trap}(f)$, in which case $Z \subset \mathrm{Trap}(f)$, or $f^2(y) < y$ for all $y \in [\varphi,\beta)$ (since $g \notin S_1$). If the latter holds then $\beta \in \Sigma_{\beta-\varphi}(f)$ and hence also $Z \subset \Sigma_{\beta-\varphi}(f)$. Therefore if we let $\nu = \min\{\eta,\varphi-\gamma,\beta-\varphi\}$ then we have

$$[\gamma,\bar{\gamma}] - (\mathrm{Trap}(f) \cup \mathrm{Homt}_2(f)) \subset \Sigma_\nu(f) ,$$

and so (5.13) holds with $\varepsilon = \nu$ and $\delta = \sigma$.

(2) and (4): If $g \notin S_1$ and g satisfies (5.12) (resp. $g \in S_6$) then by assumption (resp. by Proposition 6.2) we have that $\mathrm{Init}(g) \cup \mathrm{Trap}(g) \cup \mathrm{Homt}(g)$ is dense in $[0,1]$ and also for some $\varepsilon > 0$ $[0.1] - (\mathrm{Init}(g) \cup \mathrm{Trap}(g) \cup \mathrm{Homt}(g)) \subset \Sigma_\varepsilon(g)$. Exactly as in (1) we then have $\mathrm{Trap}(f) \cup \mathrm{Homt}(f)$ is dense in $[\gamma,\bar{\gamma}]$ and as in (3) there exists $\nu > 0$ such that

$$[\gamma,\bar{\gamma}] - (\mathrm{Trap}(f) \cup \mathrm{Homt}(f)) \subset \Sigma_\nu(f) ;$$

i.e. f satisfies (5.12). This completes the proof of Proposition 6.5. ☐☐

We now continue the proof of Proposition 6.4. Recall that the first four parts of this proposition follow directly from Proposition 6.5 and therefore we have only the last part left to consider. Let

$$S_\infty = \{ f \in S : V^n f \in S_2 \cup S_3 \text{ for all } n \geq 0 \} ;$$

our task is thus to show that if $f \in S_\infty$ then (5.14) holds.

Let $f \in S_\infty$; for each $n \geq 1$ there thus exists a reduction $r_n = ([c_n,d_n],(V^{n-1}f)^{m_n})$ of $([0,1],V^{n-1}f)$, and $V^n f$ is then a linear rescaling of the restriction of $(V^{n-1}f)^{m_n}$ to $[c_n,d_n]$. Let us rewrite this without the rescaling: define $\{a_n\}_{n\geq 1}$, $\{b_n\}_{n\geq 1}$ by $a_1 = c_1$, $b_1 = d_1$ and $a_{n+1} = (b_n-a_n)c_{n+1} + a_n$, $b_{n+1} = (b_n-a_n)d_{n+1} + a_n$; also let $q_n = \prod_{k=1}^{n} m_k$. We thus have $a_n \leqq a_{n+1} < \varphi < b_{n+1} \leqq b_n$, and the restriction of f^{q_n} to $[a_n,b_n]$ is a scaled down (or scaled down and turned upside down) version of $V^n f$. Let g_n denote the restriction of f^{q_n} to $[a_n,b_n]$ and let

$$[u_n,v_n] = \begin{cases} [g_n^2(\varphi),g_n(\varphi)] & \text{if } g_n \text{ is increasing on } [a_n,\varphi] , \\ [g_n(\varphi),g_n^2(\varphi)] & \text{if } g_n \text{ is increasing on } [\varphi,b_n] . \end{cases}$$

(See the pictures at the top of the next page.)

Lemma 6.5 $u_n < \varphi < v_n$.

Proof We only consider the case when g_n is increasing on $[a_n,\varphi]$, the proof in the other case is almost exactly the same. Now g_n is a linear

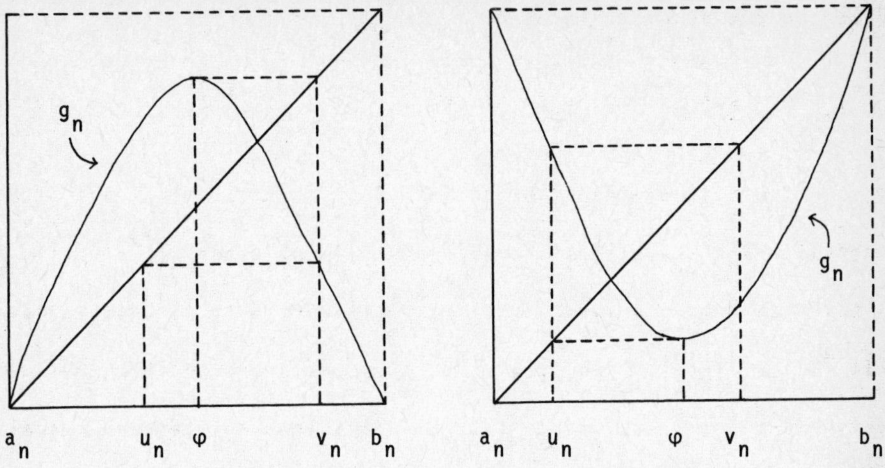

rescaling of $V^n f$ and $V^n f \in S_2 \cup S_3$; thus $g_n(\varphi) > \varphi$ and so we have $v_n = g_n(\varphi) > \varphi$. If $u_n \geqq \varphi$ then it is easy to see that $V^n f \in S_2$ and $T(V^n f) \in S_1$; but this would imply that $V^{n+1} f \in S_1$, which is not the case. Therefore $u_n < \varphi$. □□

Note that $g_n([u_n,v_n]) = [u_n,v_n]$, thus if we put $I_0^{(n)} = [u_n,v_n]$ and for $k = 1,\ldots,q_n$ let $I_k^{(n)} = f^k(I_0^{(n)})$ then $f(I_k^{(n)}) = I_{k+1}^{(n)}$ for $k = 0,\ldots,q_n-1$ and $f(I_{q_n}^{(n)}) = I_0^{(n)}$. Let $I^{(n)} = \bigcup_{k=0}^{q_n-1} I_k^{(n)}$.

Lemma 6.6 For each $n \geqq 1$ we have $I^{(n+1)} \subset I^{(n)}$.

Proof Since $g_n([u_n,v_n]) = [u_n,v_n]$ we have $g_n^k(\varphi) \in [u_n,v_n]$ for all $k \geqq 0$. But $I_0^{(n+1)}$ has $g_n^{m_{n+1}}(\varphi)$ and $g_n^{2m_{n+1}}(\varphi)$ as its end-points and so $I_0^{(n+1)} \subset I_0^{(n)}$. It thus follows that $I_{kq_n}^{(n+1)} \subset I_0^{(n)}$ for $k = 0,\ldots,m_{n+1}-1$ (because $f^{q_n}(I_0^{(n)}) = I_0^{(n)}$) and hence that

$I_k^{(n+1)} \subset I_j^{(n)}$ whenever $k \equiv j \pmod{q_n}$. □□

Lemma 6.7 For each $n \geq 1$ the intervals $I_0^{(n)},\dots,I_{q_n-1}^{(n)}$ are disjoint.

Proof We prove this by induction on n . Suppose for some $\ell \geq 1$ that the intervals $I_0^{(\ell)},\dots,I_{q_\ell-1}^{(\ell)}$ are disjoint. Then, since $I_k^{(\ell+1)} \subset I_j^{(\ell)}$ whenever $k \equiv j \pmod{q_\ell}$ and since $f^{q_{\ell+1}}(I_k^{(\ell+1)}) = I_k^{(\ell+1)}$ it follows that $I_0^{(\ell+1)},\dots,I_{q_{\ell+1}-1}^{(\ell+1)}$ would also be disjoint provided that the intervals $I_{kq_\ell}^{(\ell+1)}$, $k = 0,\dots,m_{\ell+1}-1$ are disjoint. However, on rescaling it is easy to see that this last statement holds if and only if the lemma is true in the case $n = 1$ for the function $\mathcal{V}^\ell f$. Since $\mathcal{V}^\ell f \in \mathcal{S}_\infty$ it is thus sufficient to consider the case $n = 1$. Hence we must show that if $\Gamma = ([c,d],f^m)$ is the reduction of $([0,1],f)$ given by either Proposition 6.1 or Proposition 6.3 then the intervals $[u,v]$, $f([u,v])$,..., $f^{m-1}([u,v])$ are disjoint, where

$$[u,v] = \begin{cases} [f^{2m}(\varphi),f^m(\varphi)] & \text{if } f^m \text{ is increasing on } [c,\varphi] , \\ [f^m(\varphi),f^{2m}(\varphi)] & \text{if } f^m \text{ is increasing on } [\varphi,d] . \end{cases}$$

If $f \in \mathcal{S}_2$ (and so $\Gamma = ([\alpha,\beta],f^2)$) then it is clear that $[u,v]$ and $f([u,v])$ are disjoint unless $f^4(\varphi) = \beta$. But $f^4(\varphi) = \beta$ is not possible since in this case there cannot exist a reduction of $([0,1],\mathcal{T}f)$ (and thus $\mathcal{T}f \notin \mathcal{S}_2 \cup \mathcal{S}_3$). If $f \in \mathcal{S}_3$ then in the same way we must have $f^{2m}(\varphi) \neq e_\Gamma$ (because if $f^{2m}(\varphi) = e_\Gamma$ then there cannot exist a reduction of $([0,1],\mathcal{U}f)$). But if $f^{2m}(\varphi) \neq e_\Gamma$ then, since Γ is simple, we have $f^k([u,v]) \subset f^k((c,d))$ for all $0 \leq k < m$, and hence by (6.2) the intervals $[u,v]$,..., $f^{m-1}([u,v])$ are disjoint. □□

Lemmas 6.6 and 6.7 show that $(\{I^{(n)}\}_{n\geq 1}, f)$ is an infinite register shift and by construction φ is contained in the interior of each $I^{(n)}$; also it is clear that $I^{(n)} \subset [\gamma, \overline{\gamma}]$ for each $n \geq 1$. We now consider the last part of (5.14). Let $\{a_n\}_{n\geq 1}$ and $\{b_n\}_{n\geq 1}$ be as before, put $a_0 = 0$, $b_0 = 1$, and for $n \geq 1$ let B'_{Γ_n} and C'_{Γ_n} be obtained from B_{Γ_n} and C_{Γ_n} by undoing the rescaling; thus

$$C'_{\Gamma_n} = \{ x \in [a_{n-1}, b_{n-1}] : f^{kq_n}(x) \in (a_n, b_n) \text{ for some } k \geq 0 \}$$

and $B'_{\Gamma_n} = \text{int}([a_{n-1}, b_{n-1}] - C'_{\Gamma_n})$ (where int here means with respect to the relative topology on $[a_{n-1}, b_{n-1}]$). Put $B''_{\Gamma_n} = B'_{\Gamma_n} \cap (a_{n-1}, b_{n-1})$ and $C''_{\Gamma_n} = C'_{\Gamma_n} \cap (a_{n-1}, b_{n-1})$, hence B''_{Γ_n} and C''_{Γ_n} are open subsets of $[0,1]$ and $B''_{\Gamma_n} \cup C''_{\Gamma_n}$ is dense in $[a_{n-1}, b_{n-1}]$. For $n \geq 1$ let

$$G_n = \{ x \in [0,1] : f^k(x) \in C''_{\Gamma_n} \cup \bigcup_{j=1}^{n} B''_{\Gamma_j} \text{ for some } k \geq 0 \} .$$

Lemma 6.8 G_n is a dense open subset of $[0,1]$ for each $n \geq 1$.

Proof Clearly each G_n is open. Now since $C''_{\Gamma_{n+1}} \cup B''_{\Gamma_{n+1}}$ is a dense subset of (a_n, b_n) we have

$$\{ x \in [0,1] : f^k(x) \in C''_{\Gamma_{n+1}} \cup B''_{\Gamma_{n+1}} \text{ for some } k \geq 0 \}$$

is a dense subset of $\{ x \in [0,1] : f^k(x) \in C''_{\Gamma_n}$ for some $k \geq 0 \}$ and from this it follows that G_{n+1} is a dense subset of G_n. But $G_1 = C''_{\Gamma_1} \cup B''_{\Gamma_1}$ is dense in $[0,1]$ and therefore G_n is dense for each $n \geq 1$. ◻◻

Note that we have

$$\{ x \in [0,1] : f^k(x) \in C''_{\Gamma_n} \text{ for some } k \geq 0 \}$$

$$\subset \{ x \in [0,1] : f^k(x) \in (a_n,b_n) \text{ for some } k \geq 0 \}$$

and that $(a_n,b_n) \subset (u_{n-1},v_{n-1}) \subset \mathrm{int}(I^{(n-1)})$. Thus if we let $G'_n = G_{n+1} \cap [\gamma,\bar{\gamma}]$ then

$$G'_n \subset \{ x \in [0,1] : f^k(x) \in \mathrm{int}(I^{(n)}) \text{ for some } k \geq 0 \}$$

$$\cup \{ x \in [0,1] : f^k(x) \in \bigcup_{j=1}^{n+1} B''_{\Gamma_j} \text{ for some } k \geq 0 \}$$

and by Lemma 6.7 G'_n is a dense open subset of $[\gamma,\bar{\gamma}]$. Therefore (5.14) will hold if we can show that for each $n \geq 1$

$$\{ x \in [0,1] : f^k(x) \in \bigcup_{j=1}^{n} B''_{\Gamma_j} \text{ for some } k \geq 0 \} \subset \mathrm{Init}(f) \cup \mathrm{Trap}(f) \cup \mathrm{Homt}_2(f) . \tag{6.7}$$

(6.7) is certainly true when $n = 1$: In this case we have $\{ x \in [0,1] : f^k(x) \in B''_{\Gamma_1} \text{ for some } k \geq 0 \} \subset B_{\Gamma_1}$ and Propositions 6.1 and 6.3 give us that $B_{\Gamma_1} \subset \mathrm{Init}(f) \cup \mathrm{Trap}(f) \cup \mathrm{Homt}_2(f)$. In general we will show that (6.7) holds by induction on n , so suppose (6.7) holds for some $n \geq 1$ (and for all functions in $\mathcal{S}_\infty$). Then applying this to the function $\mathcal{V}f$ we get that

$$\bigcup_{j=2}^{n+1} B''_{\Gamma_j} \subset \Psi_{\Gamma_1}^{-1}(\mathrm{Init}(\mathcal{V}f) \cup \mathrm{Trap}(\mathcal{V}f) \cup \mathrm{Homt}_2(\mathcal{V}f)) ;$$

but also $\bigcup_{j=2}^{n+1} B''_{\Gamma_j} \subset (c,d)$, and so by Lemma 6.4 we have $\bigcup_{j=2}^{n+1} B''_{\Gamma_j}$ is a subset of $\mathrm{Trap}(f) \cup \mathrm{Homt}_2(f)$. Thus

$$\{ x \in [0,1] : f^k(x) \in \bigcup_{j=1}^{n+1} B''_{\Gamma_j} \text{ for some } k \geq 0 \}$$

$$\subset \{ x \in [0,1] : f^k(x) \in B''_{\Gamma_1} \text{Trap}(f) \cup \text{Homt}_2(f) \text{ for some } k \geq 0 \}$$

$$\subset \text{Init}(f) \cup \text{Trap}(f) \cup \text{Homt}_2(f) \ ;$$

i.e. (6.7) holds with $n+1$ replacing n . Therefore (6.7) holds for all $n \geq 1$, and this completes the proof of the last part of Proposition 6.4. ∎

It remains to give the proofs of Propositions 6.1, 6.2 and 6.3.

Proof of Proposition 6.1 (1): We have $f^2([\alpha,\beta]) = [f^2(\varphi),\beta] \subset [\alpha,\beta]$; also $f((\alpha,\beta)) = (\beta,f(\varphi)]$ and hence $(\alpha,\beta) \cap f((\alpha,\beta)) = \phi$. Thus $([\alpha,\beta],f^2)$ is a reduction of $([0,1],f)$. Again let us put

$Z = \{ x \in [0,1] : f^n(x) = \beta$ for some $n \geq 0 \}$, so Z is countable. It is easy to check that we have $[0,1] - C_\Gamma = [0,\gamma) \cup (\bar{\gamma},1] \cup Z' \cup N$, where $Z' \subset Z$; thus $B_\Gamma = \text{int}([0,1]-C_\Gamma) = [0,\gamma) \cup (\bar{\gamma},1]$ and also $[0,1] - (B_\Gamma \cup C_\Gamma) \subset N \cup Z$.

(2): Let ξ be the smallest fixed point of f^2 in $[\varphi,\beta]$ and suppose that $f^2(\varphi) < \varphi$ and $\xi < \beta$. Since $f^2(\varphi) < \varphi$ there exists a unique point $v \in (\beta,1)$ with $f(v) = \varphi$ and $[\varphi,v]$ is then the largest interval containing β on which f^2 is increasing. We have $f^2(\xi) = \xi > \varphi = f(v)$ and so $f(\xi) < v$. Finally, let $z \in (f(\xi),v]$; then $f(z) \in [\varphi,\xi)$ and hence $f^2(z) < z$. Therefore $f(z) > z$, and this shows that $[\xi,f(\xi)]$ is a trap. ∎

Proof of Propositions 6.2 and 6.3 Let us fix $f \in \mathcal{S}$ with $f(\varphi) > \varphi$ and $f^2(\varphi) < \alpha$ (and so in particular $\alpha > 0$). Let

$E = \{ x \in [0,1] : f^k(x) \in [\alpha,1]$ for all $k \geq n$ for some $n \geq 0 \}$,

and let $D = [0,1] - E$, thus

$D = \{ x \in [0,1] : f^k(x) \in [0,\alpha)$ for infinitely many $k \geq 0 \}$.

Note that $f(E) \subset E$ and $f(D) \subset D$; also that $f^{-1}(E) \subset E$ and $f^{-1}(D) \subset D$.

Lemma 6.9 Let $x \in [0,1] - (\mathrm{int}(D) \cup \mathrm{int}(E))$; then for each $\delta > 0$ there exists $k \geq 0$ such that $[\alpha,\beta] \subset f^k((x-\delta,x+\delta))$.

Proof There exist $y, z \in (x-\delta,x+\delta)$ with $y \in E$ and $z \in D$. We can thus find $n \geq 0$ with $f^n(y) \geq \alpha$ and $f^n(z) < \alpha$, and hence $\alpha \in f^n((x-\delta,x+\delta))$. Therefore $\beta \in f^m((x-\delta,x+\delta))$ for all $m > n$. But since $z \in D$ there exists $k > n$ with $f^k(z) < \alpha$, and thus with $f^k((x-\delta,x+\delta)) \supset [\alpha,\beta]$. ⊞

Lemma 6.10 $[0,1] - (\mathrm{int}(D) \cup \mathrm{int}(E)) \subset \Sigma_\varepsilon(f)$, where $\varepsilon = \min\{\beta-\alpha, f(\varphi)-\beta\}$.

Proof We have $f([\alpha,\beta]) = [\beta,f(\varphi)]$ and $f^2([\alpha,\beta]) = [f^2(\varphi),\beta] \supset [\alpha,\beta]$; thus if $f^k(J) \supset [\alpha,\beta]$ for some $k \geq 0$ then $|f^n(J)| \geq \varepsilon$ for all $n \geq k$. The result therefore follows from Lemma 6.9. ⊞

Lemma 6.11 If $(\mathrm{int}(D) \cup \mathrm{int}(E)) \cap (\gamma,\bar{\gamma}) \neq \phi$ then $\mathrm{int}(D) \cup \mathrm{int}(E)$ is dense in $[0,1]$.

Proof For $n \geq 0$ let $M_n = \{ y \in [0,1] : f^n(y) \in \mathrm{int}(D) \cup \mathrm{int}(E) \}$; since $f^{-1}(D) \subset D$, $f^{-1}(E) \subset E$ we have $M_n \subset \mathrm{int}(D) \cup \mathrm{int}(E)$. Let $x \notin \mathrm{int}(D) \cup \mathrm{int}(E)$ and $\delta > 0$; by Lemma 6.9 there exists $k \geq 0$ with $(\alpha,\beta) \subset f^k((x-\delta,x+\delta))$ and thus with $(x-\delta,x+\delta) \cap M_k \neq \phi$, since it is easy to see that if $(\mathrm{int}(D)\cup\mathrm{int}(E))\cap(\gamma,\bar{\gamma}) \neq \phi$ then also $(\mathrm{int}(D)\cup\mathrm{int}(E))\cap(\alpha,\beta) \neq \phi$. Hence $(x-\delta,x+\delta)\cap(\mathrm{int}(D)\cup\mathrm{int}(E)) \neq \phi$ and therefore $\mathrm{int}(D)\cup\mathrm{int}(E)$ is dense in $[0,1]$. ▫▫

Lemma 6.12 Either (6.4) holds or $\mathrm{int}(D)\cup\mathrm{int}(E)$ is dense in $[0,1]$.

Proof Suppose $\mathrm{int}(D)\cup\mathrm{int}(E)$ is not dense in $[0,1]$; then, since $[0,\gamma)\cup(\bar{\gamma},1] \subset \mathrm{int}(D)$, we have by Lemma 6.11 that $(\mathrm{int}(D)\cup\mathrm{int}(E))\cap(\gamma,\bar{\gamma})$ is empty, and so by Lemma 6.10 $[\gamma,\bar{\gamma}] \subset \Sigma_\varepsilon(f)$ with $\varepsilon = \min\{\beta-\alpha, f(\varphi)-\beta\}$. But $[0,\gamma)\cup(\bar{\gamma},1] \subset \mathrm{Init}(f)$ is always true and hence Proposition 5.10 implies that $\mathrm{Trap}(f)\cup\mathrm{Homt}(f) = \phi$. Thus (6.4) holds. ▫▫

We now examine what happens when $\mathrm{int}(D)\cup\mathrm{int}(E)$ is dense in $[0,1]$. Let

$$F = \{ x \in [0,1] : f^n(x) \in [\alpha,1] \text{ for all } n \geq 0 \} ,$$

thus F is closed, $f(F) \subset F$, and of course

$$E = \{ x \in [0,1] : f^k(x) \in F \text{ for some } k \geq 0 \} .$$

Let us write $\mathrm{int}(D) = \bigcup_{s\in S} D_s$ and $\mathrm{int}(F) = \bigcup_{t\in T} F_t$ as countable disjoint unions of non-empty open intervals (with $S = \phi$ (resp. $T = \phi$) if $\mathrm{int}(D) = \phi$ (resp. $\mathrm{int}(F) = \phi$)). If $s \in S$ and $n \geq 0$ then $f^n(D_s)$ is a connected subset of D and so there exists a unique $s(n) \in S$ such that $f^n(D_s) \subset \bar{D}_{s(n)}$. Similarly, if $t \in T$ and $n \geq 0$ then there exists a unique $t(n) \in T$ with $f^n(F_t) \subset \bar{F}_{t(n)}$. It is easy to see that

(6.8) $(s(m))(n) = s(n+m)$ for all $s \in S$, $m, n \geq 0$,

and

(6.9) $(t(m))(n) = t(n+m)$ for all $t \in T$, $m, n \geq 0$.

We say that $s \in S$ (resp. $t \in T$) is *periodic* if $f^n(D_s) \subset \overline{D}_s$ (resp. $f^n(F_t) \subset \overline{F}_t$) for some $n \geq 1$; the smallest such $n \geq 1$ is called the *period* of s (resp. t). We call $s \in S$ (resp. $t \in T$) *eventually periodic* if $s(n)$ (resp. $t(n)$) is periodic for some $n \geq 0$; let S_0 (resp. T_0) denote the set of eventually periodic elements of S (resp. T).

The facts concerning $\{D_s\}_{s \in S}$ and $\{F_t\}_{t \in T}$ which will be used to prove Propositions 6.2 and 6.3 are collected together in the next result.

Proposition 6.6 (6.10) If $t \in T$ is periodic then $\overline{F}_t$ is a trap.

(6.11) If $t \in T-T_0$ then $\overline{F}_t$ is a type 2 homterval.

(6.12) If $[\alpha,\beta] \not\subset E$ then

$$\text{int}(E) \subset \{ x \in [0,1] : f^k(x) \in \bigcup_{t \in T} \overline{F}_t \text{ for some } k \geq 0 \} .$$

(6.13) If $[\alpha,\beta] \subset E$ then f satisfies (6.3).

(6.14) If $s \in S$ is periodic with period 1 then $\gamma > 0$ and $D_s = [0,\gamma)$.

(6.15) If $s \in S$ is periodic with period $n > 1$ and if $\varphi \notin D_{s(k)}$ for all $0 \leq k < n$ then $\overline{D}_s$ is a trap.

(6.16) If $s \in S$ is periodic with period $m > 1$ and if $\varphi \in D_s$ then $(\overline{D}_s, f^m)$ is a simple reduction of $([0,1],f)$.

(6.17) If $s \in S-S_0$ then $D_s \subset \text{Homt}(f)$; if also $\varphi \in \text{int}(D)$ then $D_s \subset \text{Homt}_2(f)$.

Before proving Proposition 6.6 let us show how it implies

Propositions 6.2 and 6.3. First note that from (6.10), (6.11) and (6.12) we have

(6.18) If $[\alpha,\beta] \not\subset E$ then $\mathrm{int}(E) \subset \mathrm{Trap}(f) \cup \mathrm{Homt}_2(f)$.

Proof of Proposition 6.2 By assumption we have $f^2(\varphi) < \gamma$ and hence $\varphi \in \mathrm{int}(D)$ (because $[0,\gamma) \subset \mathrm{int}(D)$). Thus by Lemma 6.1 $\mathrm{int}(D) \cup \mathrm{int}(E)$ is dense in $[0,1]$. Now if $s \in S$ is periodic with period $n > 1$ then clearly $\varphi \notin D_{s(k)}$ for all $0 \leq k < n$, and therefore (6.14), (6.15) and (6.17) imply that $\mathrm{int}(D) \subset \mathrm{Init}(f) \cup \mathrm{Trap}(f) \cup \mathrm{Homt}_2(f)$. Together with (6.18) this gives us

$$\mathrm{int}(D) \cup \mathrm{int}(E) \subset \mathrm{Init}(f) \cup \mathrm{Trap}(f) \cup \mathrm{Homt}_2(f) .$$

Thus $\mathrm{Init}(f) \cup \mathrm{Trap}(f) \cup \mathrm{Homt}_2(f)$ is dense in $[0,1]$, and by Lemma 6.10 we have

$$[0,1] - (\mathrm{Init}(f) \cup \mathrm{Trap}(f) \cup \mathrm{Homt}_2(f))$$
$$\subset [0,1] - (\mathrm{int}(D) \cup \mathrm{int}(E)) \subset \Sigma_\varepsilon(f)$$

(with $\varepsilon = \min\{\beta-\alpha, f(\varphi)-\beta\}$). ▫▫

Proof of Proposition 6.3 We are assuming that $\gamma \leq f^2(\varphi) < \alpha$ and so we clearly have $\mathrm{Init}(f) = [0,\gamma) \cup (\bar{\gamma},1]$. If $[\alpha,\beta] \subset E$ then by (6.13) we know that (6.3) holds; we can therefore assume that $[\alpha,\beta] \not\subset E$. Suppose first that φ does not belong to any D_s with s periodic; then (6.14), (6.15) and (6.16) imply that $\mathrm{int}(D) \cap [\gamma,\bar{\gamma}] \subset \mathrm{Trap}(f) \cup \mathrm{Homt}(f)$, and thus (with the help of (6.18)) we have

$$(\mathrm{int}(D) \cup \mathrm{int}(E)) \cap [\gamma,\bar{\gamma}] \subset \mathrm{Trap}(f) \cup \mathrm{Homt}(f) .$$

Lemmas 6.10 and 6.12 now tell us that either (6.3) or (6.4) holds.

This leaves us with the case when $\varphi \in D_s$ for some periodic $s \in S$ with period $m > 1$. By (6.16) $\Gamma = (\bar{D}_s, f^m)$ is then a simple reduction of

$([0,1],f)$, and it is easy to see that $C_\Gamma = \bigcup_{u\in V_s} D_u$, where $V_s = \{ u \in S : u(n) = s$ for some $n \geqq 0 \}$. (*Note:* Let $u \in V_s$ and let $k \geqq 0$ be the smallest integer with $u(k) = s$; then we have $f^k(D_u) \subset D_s$ (rather than just $f^k(D_u) \subset \overline{D}_s$), and this shows that $D_u \subset C_\Gamma$.) Hence $C_\Gamma \subset \mathrm{int}(D)$ and also $\mathrm{int}(D)-C_\Gamma$ is open, and it thus follows that $B_\Gamma \cup C_\Gamma \supset \mathrm{int}(D)\cup\mathrm{int}(E)$. Lemma 6.10 therefore gives us that $[0,1] - (B_\Gamma \cup C_\Gamma) \subset \Sigma_\varepsilon(f)$ (again with $\varepsilon = \min\{\beta-\alpha, f(\varphi)-\beta\}$). Finally, Lemma 6.9 implies that $B_\Gamma \subset \mathrm{int}(D)\cup\mathrm{int}(E)$, and so by (6.14), (6.15), (6.17) and (6.18) we have $B_\Gamma \subset \mathrm{Init}(f)\cup\mathrm{Trap}(f)\cup\mathrm{Homt}_2(f)$. Thus in this case (6.5) holds. ◻◻

Proof of Proposition 6.6 Note that $\varphi \notin F$ (since $f^2(\varphi) < \alpha$) and that $F \subset [\alpha,1)$. ($1 \notin F$ because $f(1) \leqq f^2(\varphi) < \alpha$.) We first deal with (6.10).

Lemma 6.13 If $t \in T$ is periodic then $\overline{F}_t$ is a trap.

Proof Put $\overline{F}_t = [a,b]$; since F is closed we have $[a,b] \subset F$ and so in particular $[a,b] \subset (0,1)$. $[a,b]$ is a sink (because $\varphi \notin F$), thus let m be the smallest positive integer such that both f^m is increasing on $[a,b]$ and $f^m([a,b]) \subset [a,b]$. (Clearly m is either n or $2n$, where n is the period of t .) Suppose that $f^m(b) < b$; then we can find $z > b$ with $z \notin F$ and $f^m([b,z]) \subset [a,b]$. Since $z \notin F$ (and because $f^j(z) \in F \subset [\alpha,1]$ for all $j \geqq m$) there exists $0 \leqq k < m$ with $f^k(z) < \alpha$. But $f^k(b) \geqq \alpha$ and so $f^k(y) = \alpha$ for some $y \in [b,z]$. It is now easy to see that $[b,y] \subset F$, thus $b = y$ and hence α is an end-point of $\overline{F}_{t(k)}$. This is clearly not possible and therefore $f^m(b) = b$; in the same way we obtain $f^m(a) = a$. Let $[u,v]$ be the largest interval containing $[a,b]$ on which f^m is increasing. We have $u < a < b < v$ (because if a or b was a turning point of f^m then we would have $\varphi \in F$). If $f^m(y) \leq y$ for some $y \in (b,v]$ then a similar proof to the above would show that α is an end-point of $\overline{F}_{t(k)}$ for some $0 \leq k < m$. Thus $f^m(y) > y$ for all $y \in (b,v]$; in the same way $f^m(y) < y$ for all $y \in [u,a)$, and so $\overline{F}_t$ is a trap. ◻◻

Lemma 6.13 gives us (6.10); we next consider (6.12) and (6.13). ((6.11) will be dealt with at the end together with (6.17).)

Lemma 6.14 If $[\alpha,\beta] \not\subset E$ then

$$\text{int}(E) \subset \{ x \in [0,1] : f^k(x) \in \bigcup_{t\in T} \overline{F}_t \text{ for some } k \geq 0 \} .$$

Proof Let $x \in \text{int}(E)$, there thus exists an open interval $J \subset \text{int}(J)$ with $x \in J$; since $x \in E$ we can also find $m \geq 0$ such that $f^k(x) \in [\alpha,1]$ for all $k \geq m$. If $\alpha \notin f^k(J)$ for all $k \geq 0$ then $f^k(J) \subset [\alpha,1]$ for all $k \geq m$ and thus $f^m(J) \subset \overline{F}_t$ for some $t \in T$; in particular $f^m(x) \in \overline{F}_t$ for some $t \in T$. Assume then that $\alpha \in f^n(J)$ for some $n \geq 0$; hence $\beta \in f^{n+1}(J)$. If $f^{n+1}(J) \subset F$ then we again have $f^{n+1}(x) \in \overline{F}_t$ for some $t \in T$, and so we are left with the case when $f^{n+1}(J) \not\subset F$. If this holds then there exists $y \in J$ and $k \geq n+1$ with $f^k(y) < \alpha$; but $\beta \in f^k(J)$ and hence in this case we have $[\alpha,\beta] \subset f^k(J) \subset E$. ◻◻

Lemma 6.15 Let ξ be the smallest fixed point of f^2 in $[\varphi,\beta]$ and suppose that $f^3(\varphi) < \xi$. Then $[\alpha,\beta] \not\subset E$.

Proof We have $f^2(\varphi) < \alpha$ and $f^2(\xi) = \xi > \alpha$; there thus exists $\eta \in (\varphi,\xi)$ with $f^2(\eta) = \alpha$. Similarly we have $f^3(\varphi) < \xi$ and $f^3(\eta) = f(\alpha) = \beta \geq \xi$, and so there exists $\nu \in (\varphi,\eta] \subset (\varphi,\xi)$ with

$f^3(\nu) = \xi$.

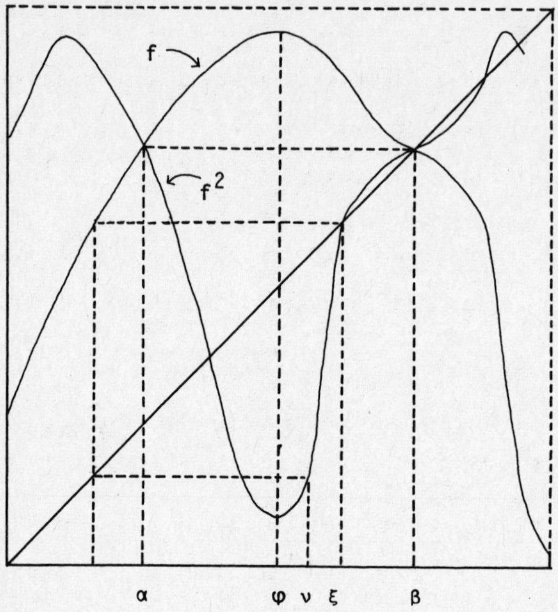

Choose $\delta > 0$ so that $\nu-\delta > \varphi$ and let $J = f^3([\nu-\delta,\nu])$; J is then a closed interval having ξ as its right-hand end-point. Now f^2 is increasing on $[\varphi,\xi]$, $f^2(\xi) = \xi$ and $f^2(z) < z$ for all $z \in [\varphi,\xi)$; we can therefore find a closed interval $I \subset J$ and $n \geq 1$ such that $f^{2n}(I) = [\nu-\delta,\nu]$, and this gives us that $f^{2n+3}([\nu-\delta,\nu]) \supset [\nu-\delta,\nu]$. Put $m = 2n+3$ and let

$$Y = \{ x \in [\nu-\delta,\nu] : f^{mk}(x) \in [\nu-\delta,\nu] \text{ for all } k \geq 0 \} ;$$

by compactness we have $Y \neq \phi$. But if $y \in Y$ then $f^{mk+1}(y) \in f([\nu-\delta,\nu]) \subset [0,\alpha]$ for all $k \geq 0$, and thus in fact $f^{mk+1}(y) \in [0,\alpha)$ for all $k \geq 0$ (since if $f^{mk+1}(y) = \alpha$ then

$f^j(y) = \beta$ for all $j \geq mk+2$, and this is not possible if $y \in Y$). Hence $Y \subset D$, and so in particular $[\alpha,\beta] \not\subset E$. ▫▫

Lemma 6.16 If $[\alpha,\beta] \subset E$ then $\mathrm{Trap}(f) \cup \mathrm{Homt}_2(f)$ is dense in $[\gamma,\bar{\gamma}]$.

Proof We have $[\alpha,\beta] \subset E$ and so by Lemma 6.15 $\xi \leq f^3(\varphi) < \beta$. ($f^3(\varphi) < \beta$ follows because $f^2(\varphi) < \alpha$.) As in the proof of Proposition 6.1(2) $[\xi,f(\xi)]$ is a trap and it is easy to see that there exists a periodic $u \in T$ with period 1 such that $F_u = (\xi,f(\xi))$. Now if $W = \{ x \in [\alpha,\beta] : f^2(x) < \alpha \}$ then $f^3(W) \subset [\xi,\beta] \subset \bar{F}_u$, and hence

$$[\alpha,\beta] - F \subset \{ x \in [0,1] : f^k(x) \in \bar{F}_u \text{ for some } k \geq 0 \} .$$

But $[\alpha,\beta] - F$ is open and so (6.10) implies that $[\alpha,\beta] - F \subset \mathrm{Trap}(f)$; therefore by (6.10) and (6.11) we have $[\alpha,\beta] - (\mathrm{Trap}(f) \cup \mathrm{Homt}_2(f)) \subset V$, where $V = \{ x \in [0,1] : f^k(x) \in \partial F \text{ for some } k \geq 0 \}$. From this it immediately follows that $[\gamma,\bar{\gamma}] - (\mathrm{Trap}(f) \cup \mathrm{Homt}_2(f)) \subset V \cup N$, where $N \subset \{\gamma,\bar{\gamma}\}$ is as in Proposition 6.1(1). Now $\{ x \in [0,1] : f^k(x) \in \partial F \}$ is nowhere dense for each $k \geq 0$ (because ∂F is nowhere dense and $|\{ z : f^k(z) = y \}| \leq 2^k$ for each $y \in [0,1]$), and hence by the Baire category theorem V is also nowhere dense. Thus $\mathrm{Trap}(f) \cup \mathrm{Homt}_2(f)$ is dense in $[\gamma,\bar{\gamma}]$. ▫▫

Remark: In the above proof we made use of (6.11) (which of course has not yet been proved); however, the proof of (6.11) is independent of what we have just done, and so this causes no problems.

Lemma 6.17 If $[\alpha,\beta] \subset E$ then $[\gamma,\bar{\gamma}] - (\mathrm{Trap}(f) \cup \mathrm{Homt}_2(f)) \subset \Sigma_\varepsilon(f)$, where $\varepsilon = \min\{\beta-\alpha, f(\varphi)-\beta\}$.

Proof We continue to use the notation of Lemma 6.16. Let $J \subset [\gamma,\bar{\gamma}]$ be a non-empty open interval. If either $f^m(J) \subset [\xi,f(\xi)]$ for some $m \geq 0$ or $f^k(J) \cap [\xi,f(\xi)] = \phi$ for all $k \geq 0$ then clearly $f^n(J) \subset F$ for some $n \geq 0$, and hence by (6.10) and (6.11) we have $J \subset \mathrm{Trap}(f) \cup \mathrm{Homt}_2(f)$. If these two alternatives do not hold then for some $k \geq 0$ we have both $f^k(J) \cap [\xi,f(\xi)] \neq \phi$ and $f^k(J) \not\subset [\xi,f(\xi)]$. In this case it is easy to see that there exists $m \geq k$ with $f^m(J) \supset [\alpha,\xi]$, and thus $|f^n(J)| \geq \varepsilon$ for all $n \geq m+2$ because $f([\alpha,\xi]) = [\beta,f(\varphi)]$ and $f([\beta,f(\varphi)]) \supset [\alpha,\beta]$. This shows that $[\gamma,\bar{\gamma}] - N \subset \mathrm{Trap}(f) \cup \mathrm{Homt}_2(f) \cup \Sigma_\varepsilon(f)$. A similar argument gives us $N \subset \Sigma_\varepsilon(f)$ and therefore $[\gamma,\bar{\gamma}] - (\mathrm{Trap}(f) \cup \mathrm{Homt}_2(f)) \subset \Sigma_\varepsilon(f)$. ▫▫

Lemmas 6.14, 6.16 and 6.17 give us (6.12) and (6.13).

We now consider the intervals $\{D_s\}_{s \in S}$. Suppose $s \in S$ is periodic with period 1 and put $\bar{D}_s = [a,b]$. Then $f([a,b]) \subset [a,b]$ and thus $a < \alpha$; this is only possible if $[a,b] \subset [0,\gamma]$. In this case we have $D_s = [0,\gamma)$ because $[0,\gamma) \subset D$. Therefore (6.14) holds.

Lemma 6.18 Let $s \in S$ and let u be an end-point of $\bar{D}_s$. If $u \in (0,1)$ then $f^k(u)$ is an end-point of $\bar{D}_{s(k)}$ for each $k \geq 0$.

Proof Suppose that $f^k(u)$ is not an end-point of $\bar{D}_{s(k)}$ for some $k \geq 0$; since $f^k(u) \in f^k(\bar{D}_s) \subset \bar{D}_{s(k)}$ we then have $f^k(u) \in D_{s(k)}$. But this would imply that $u \in \mathrm{int}(D)$ (because $f^{-1}(D) \subset D$), which is clearly not true. ▫▫

Lemma 6.19 Let $s \in S$ be periodic with period $n > 1$ and suppose that $\varphi \notin D_{s(k)}$ for all $0 \leq k < n$; put $\bar{D}_s = [a,b]$. Then $[a,b] \subset (0,1)$ and $f^m(a) = a$, $f^m(b) = b$, where $m = n$ (resp. $m = 2n$) if f^n is increasing (resp. decreasing) on $[a,b]$.

Proof We have f^m is increasing on $[a,b]$ and $f^m([a,b]) \subset [a,b]$. Thus if $[a,b] \subset (0,1)$ then Lemma 6.18 gives us that $f^m(a) = a$ and $f^m(b) = b$. Now $f([0,1]) \subset [0,f(\varphi)]$ and $f^n([a,b]) \subset [a,b]$; hence $a < f(\varphi)$. Therefore $b \neq 1$, since if $b = 1$ then we would have $f(\varphi) \in D_s$ and this would imply that $\varphi \in D_{s(n-1)}$. Similarly we have $f^n([a,b]) \subset [f^2(\varphi),1]$ (because $n > 1$) and hence $f^2(\varphi) < b$. Therefore $a \neq 0$, since if $a = 0$ then we would have $f^2(\varphi) \in D_s$ and thus $\varphi \in D_{s(n-2)}$. ▫▫

Lemma 6.20 φ is not an end-point of any of the intervals $\overline{D}_s$, $s \in S$.

Proof Suppose that φ is an end-point of some $\overline{D}_s$. Then, since $f^{-1}(f(D_s)) \subset D$, we have $(\varphi-\delta,\varphi)\cup(\varphi,\varphi+\delta) \subset D$ for some $\delta > 0$ and hence $\varphi \in E$.. It follows that $\alpha \in f^k((\varphi-\delta,\varphi)\cup(\varphi,\varphi+\delta))$ for some $k \geq 0$ and thus $\alpha \in f^k(D) \subset D$. But this is not possible because $\alpha \in E$. ▫▫

Lemma 6.21 Let $s \in S$ be periodic with period $n > 1$ and suppose that $\varphi \notin D_{s(k)}$ for all $0 \leq k < n$; then $\overline{D}_s$ is a trap.

Proof Put $\overline{D}_s = [a,b]$ and let $m = n$ (resp. $m = 2n$) if f^n is increasing (resp. decreasing) on $[a,b]$. f^m is increasing on $[a,b]$, and by Lemma 6.19 we have $f^m(a) = a$, $f^m(b) = b$ and $[a,b] \subset (0,1)$. Let $[u,v]$ be the largest interval containing $[a,b]$ on which f^m is increasing; we have $u < a < b < v$. (If either $u = a$ or $b = v$ then by Lemma 6.18 φ would be an end-point of $\overline{D}_{s(k)}$ for some $k \geq 0$, and by Lemma 6.20 this is not possible.) Now $b \notin \mathrm{int}(D)\cup\mathrm{int}(E)$, and so Lemma 6.9 implies that for each $\delta > 0$ there exists $k \geq 0$ such that $f^k((b-\delta,b+\delta)) \supset [\alpha,\beta]$. Hence for each $\delta > 0$ there exists $k \geq 0$ and

$y \in (b,b+\delta)$ with $f^k(y) = \varphi$. Clearly this can only happen if $f^m(z) > z$ for all $z \in (b,v]$. Similarly we have $f^m(z) < z$ for all $z \in [u,a)$ and therefore $[a,b]$ is a trap. ⊞

Lemma 6.22 Let $s \in S$ be periodic with period $m > 1$ and suppose that $\varphi \in D_s$; put $\overline{D}_s = [c,d]$. Then $([c,d],f^m)$ is a simple reduction of $([0,1],f)$.

Proof It is clear that (6.1) and (6.2) hold and hence $\Gamma = ([c,d],f^m)$ is a reduction of $([0,1],f)$. We have $[c,d] \subset [\alpha,\beta]$ (because α and β are in E) and so $[c,d] \subset (0,1)$; Lemma 6.18 thus implies that $f^m(c)$ and $f^m(d)$ are either c or d and therefore we have $f^m(c) = f^m(d) = e_\Gamma$. Now without loss of generality assume that $e_\Gamma = c$ and let $[u,\varphi]$ be the largest interval containing c on which f^m is increasing; from Lemmas 6.1 and 6.2 we know that $u < c$. Suppose that $f^m(y) \geq y$ for some $y \in [u,c)$; then $f^m([y,c]) \subset [y,c]$ and so $f^m([y,d]) \subset [y,d]$. But $[y,c] \cap E \neq \phi$ and $(c,d) \subset D$; it thus follows that $\alpha \in f^k([y,c])$ for some $k \geq 0$, and hence also $\beta \in f^m(f^k([y,c])) \subset f^k([y,c])$. This implies that $f^k([y,c]) \supset [\alpha,\beta]$, which is not possible. Therefore $f^m(z) < z$ for all $z \in [u,c)$ and so Γ is a simple reduction of $([0,1],f)$. ⊞

Lemmas 6.21 and 6.22 give us (6.15) and (6.16); we thus have (6.11) and (6.17) left to check.

Lemma 6.23 If $s \in S-S_0$ then the intervals $\{D_{s(n)}\}_{n \geq 0}$ are disjoint.

Proof Let $s \in S$ and suppose there exist $0 \leq m < n$ such that $D_{s(m)} \cap D_{s(n)} \neq \phi$. Since $D_{s(m)} \cup D_{s(n)} \subset \mathrm{int}(D)$ we must then in fact have $s(m) = s(n)$. Thus $s(m)$ is periodic and hence $s \in S_0$. ⊞

Lemma 6.24 If $s \in S$ (resp. $t \in T$) and there exists a sink J and $n \geq 0$ with $f^n(\bar{D}_s) \subset J$ (resp. $f^n(\bar{F}_t) \subset J$) then $s \in S_0$ (resp. $t \in T_0$).

Proof Let $s \in S$, $n \geq 0$ and J be a sink with $f^n(\bar{D}_s) \subset J$. If $\mathrm{int}(J) \not\subset D$ then $\mathrm{int}(J) \not\subset \mathrm{int}(D) \cup \mathrm{int}(E)$ and so by Lemma 6.9 there would exist $k \geq 0$ with $f^k(\mathrm{int}(J)) \supset [\alpha,\beta]$. But this is not possible and thus $\mathrm{int}(J) \subset D$. Let $u \in S$ be such that $\mathrm{int}(J) \subset D_u$; since J is a sink u is periodic. However, we clearly have $s(n) = u$ and hence $s \in S_0$. Next let $t \in T$ with $f^n(\bar{F}_t) \subset J$, where again J is a sink and $n \geq 0$. If $J \not\subset F$ then, since $J \cap F \neq \phi$, we would have $\alpha \in f^k(J)$ for some $k \geq 0$, thus $\beta \in f^m(J)$ for all $m > k$ and therefore $\beta \in J$. But this would imply that $f^m(J) \subset [\varphi,1]$ for all $m \geq 0$ and thus that $J \subset F$. Hence we must have $J \subset F$ and as in the first part this gives us that $t \in T_0$. ⊞

Now let $t \in T\text{-}T_0$; we have $\varphi \notin f^n(\bar{F}_t)$ for all $n \geq 0$ (because $f^n(\bar{F}_t) \subset F$ and $\varphi \notin F$) and this, together with Lemma 6.24, shows that $\bar{F}_t$ is a homterval. In fact $\bar{F}_t$ must be a type 2 homterval since there exists $\delta > 0$ with $(\varphi-\delta,\varphi+\delta) \cap F = \phi$. If $s \in S\text{-}S_0$ then by Lemma 6.23 we can find $n \geq 0$ such that $\varphi \notin \bar{D}_{s(k)}$ for all $k \geq n$, and so again using Lemma 6.24 we have that $\bar{D}_{s(n)}$ is a homterval; hence $D_s \subset \mathrm{Homt}(f)$. Finally, if in addition we have $\varphi \in \mathrm{int}(D)$ then there exists $m \geq 0$ and $\delta > 0$ such that $(\varphi-\delta,\varphi+\delta) \cap \bar{D}_{s(k)} = \phi$ for all $k \geq m$; in this case we have $D_s \subset \mathrm{Homt}_2(f)$.

This completes the proof of Proposition 6.6 and thus also the proofs of Propositions 6.2 and 6.3. ⊞

Notes: The idea of a reduction is taken from Guckenheimer (1979): If $\Gamma = ([c,d],f^m)$ is a simple reduction of $([0,1],f)$ then e_Γ is what

Guckenheimer calls a restrictive central fixed point of f^m . A similar idea also occurs in Jonker and Rand (1981). The mapping $T : S_2 \to S$ (where $Tf = U_\Gamma f$ and $\Gamma = ([\alpha,\beta],f^2)$ is the reduction of $([0,1],f)$ given by Proposition 6.1) is a version of the "doubling transformation" due to Feigenbaum (Feigenbaum (1978), (1979)). We have only used some very superficial properties of this transformation; its application to the study of one-parameter families of functions can be found for example in Collet and Eckmann (1980) or Collet, Eckmann and Lanford (1980).

7. GETTING RID OF HOMTERVALS

In this section we prove the first two parts of Proposition 5.9, namely:

(1) If $f \in S$ is such that

(7.1) f has a continuous second derivative in $(0,\varphi)\cup(\varphi,1)$ and $f'(z) \neq 0$ for all $z \in (0,\varphi)\cup(\varphi,1)$

then f has no type 2 homtervals.

(2) Suppose $f \in S_R$ satisfies

(7.2) f has a continuous derivative in $(0,\varphi)\cup(\varphi,1)$ and $f'(z) \neq 0$ for all $z \in (0,\varphi)\cup(\varphi,1)$,

(7.3) $\limsup_{z\uparrow\varphi} \left|\frac{f'(z)}{f'(\underline{z})}\right| < +\infty$ and $\liminf_{z\uparrow\varphi} \left|\frac{f'(z)}{f'(\underline{z})}\right| > 0$, where $\underline{z} \neq z$ is such that $f(\underline{z}) = f(z)$;

then f has no type 1 homtervals.

Recall that if $f \in S$ and $J \subset [0,1]$ is a non-trivial closed interval then J is called a *homterval* of f if

(7.4) $\varphi \notin f^n(J)$ for all $n \geq 0$,

(7.5) $f^n(J)$ is not contained in any sink for each $n \geq 0$.

If J is a homterval of f then by Proposition 5.8(1) we know that

(7.6) the intervals $\{f^n(J)\}_{n\geq 0}$ are disjoint.

Recall further that if J is a homterval of f and $x \in J$ then J is is called a *type 1* (resp. *type 2*) homterval if $\liminf_{n\to\infty} |f^n(x)-\varphi| = 0$ (resp. $\liminf_{n\to\infty} |f^n(x)-\varphi| > 0$). (These definitions do not depend on the choice of $x \in J$ because by (7.6) we have $\lim_{n\to\infty} |f^n(J)| = 0$.)

We start the proof of Proposition 5.9 by somewhat repeating part of the proof of Theorem 3.1: Consider $f \in \mathcal{S}$ to be fixed and let $H = [0,1] - \overline{M}$, where

$$M = \{ x \in [0,1] : f^n(x) = \varphi \text{ for some } n \geq 0 \} .$$

H is open and so we can write $H = \bigcup_{t\in V} H_t$ as a countable disjoint union of non-empty open intervals (with $V = \phi$ if $H = \phi$). As noted in Section 3 the H_t are exactly the maximal open intervals J with the property that $\varphi \notin f^n(J)$ for all $n \geq 0$. Now we have $f(H) \subset H$, thus if $t \in V$ and $n \geq 0$ then there exists a unique $t(n) \in V$ such that $f^n(H_t) \subset H_{t(n)}$. Again as in Section 3 we call $t \in V$ *periodic* if $t(n) = t$ for some $n \geq 1$, and *eventually periodic* if $t(m)$ is periodic for some $m \geq 0$. Let V_0 denote the set of eventually periodic elements of V .

Proposition 7.1 If J is a homterval of f then $\text{int}(J) \subset H_t$ for some $t \in V-V_0$.

Proof Let $J \subset [0,1]$ be a non-trivial closed interval. If (7.4) holds then clearly $\text{int}(J) \subset H$; however, if $\text{int}(J) \subset H_t$ for some $t \in V_0$ then J is not a homterval (since $f^m(J) \subset \overline{H}_{t(m)}$ for all $m \geq 0$ and because if $s \in V$ is periodic then $\overline{H}_s$ is a sink). ◻◻

The proof of Proposition 5.9(1) is now really the same as the proof of Proposition 3.4: Let $f \in \mathcal{S}$ satisfy (7.1) and suppose there exists a type 2 homterval J of f ; thus $\liminf_{n\to\infty} |f^n(x)-\varphi| > 0$ for all $x \in J$, and by Proposition 7.1 there exists $t \in V-V_0$ with

$int(J) \subset H_t$. Since $t \in V-V_0$ we clearly have $t(n) \neq t(m)$ whenever $n \neq m$; the intervals $\{H_{t(n)}\}_{n\geq 0}$ are therefore disjoint and so in particular $\lim_{n\to\infty} |H_{t(n)}| = 0$. But if $x \in int(J)$ then $f^n(x) \in H_{t(n)}$, and hence we can find $\eta > 0$ and $k \geq 0$ so that $H_{t(n)} \cap (\varphi-\eta,\varphi+\eta) = \phi$ for all $n \geq k$. From this it easily follows that there exists $\delta > 0$ and $m \geq k$ such that $H_{t(n)} \subset I(\delta) = [\delta,\varphi-\delta] \cup [\varphi+\delta,1-\delta]$ for all $n \geq m$. Let $\overline{H}_{t(m)} = [c,d]$; Lemmas 3.6 and 3.7 now show that there exists $\varepsilon > 0$ with $f^n((c,d+\varepsilon)) \subset I(\delta/2)$ for all $n \geq 0$. However, this implies in particular that $(c,d+\varepsilon) \subset H$, which is not the case. Therefore if $f \in S$ satisfies (7.1) then f can have no type 2 homtervals.

The rest of this section is taken up with the proof of Proposition 5.9(2). We start with a simple observation which will be needed several times later.

Lemma 7.1 If $f \in S$ is central then f has no type 1 homtervals.

Proof This is clear. ◻◻

Now let us fix $f \in S_R$ satisfying (7.2) and (7.3) and suppose there exists a type 1 homterval J of f ; thus $\liminf_{n\to\infty} |f^n(x)-\varphi| = 0$ for all $x \in J$ and by Proposition 7.1 there exists $t \in V-V_0$ with $int(J) \subset H_t$. The intervals $\{H_{t(n)}\}_{n\geq 0}$ are again disjoint, and so we can find $m \geq 0$ such that both $H_{t(m)} \subset (0,1)$ and $\varphi \notin \overline{H}_{t(n)}$ for all $n \geq m$. Put $U = H_{t(m)}$ and for each $n \geq 0$ let $U_n = f^n(U)$. Since $U_n \subset H_{t(n+m)}$ we have

(7.7) the intervals $\{U_n\}_{n\geq 0}$ are disjoint,

(7.8) $\varphi \notin \overline{U}_n$ for all $n \geq 0$.

We clearly also have

(7.9) $U \subset (0,1)$,

(7.10) $\liminf_{n\to\infty} |f^n(x)-\varphi| = 0$ for all $x \in U$.

The proof of Proposition 5.9(2) will consist in showing that $\limsup_{n\to\infty} |U_n| > 0$, which of course is in contradiction to (7.7).

If I and J are sub-intervals of $[0,1]$ then we say that I *is higher than* J (relative to f) if $f(x) > f(y)$ for all $x \in I$, $y \in J$.

Lemma 7.2 (1) If $m \neq n$ then either U_m is higher than U_n or U_n is higher than U_m .

(2) For each $n \geq 0$ there exists $m > n$ such that U_m is higher than U_n .

Proof (1): Suppose neither held; there then exists $x \in U_n$ and $y \in U_m$ with $f(x) = f(y)$. But this would imply that $U_{n+1} \cap U_{m+1} \neq \phi$, which by (7.7) is not the case.

(2): If for some $n \geq 0$ there did not exist $m > n$ such that U_m is higher than U_n then by (1) U_n is higher than U_m for all $m > n$; but this is not possible because it clearly contradicts (7.10). ⊞

Lemma 7.2(2) allows us to inductively define a sequence of integers $\{m_n\}_{n\geq 0}$ by letting $m_0 = 0$ and

$$m_{n+1} = \min\{\ k > m_n :\ U_k \text{ is higher than } U_{m_n}\ \} .$$

Our aim is now to show that $\liminf_{n\to\infty} |U_{m_n}| > 0$, which will give us our required contradiction.

For $n \geq 1$ let

$$W_n = \{ x \in [0,1] : f^k(x) < f(x) \text{ for } k = 2,\dots,n \text{ and } f^{n+1}(x) > f(x) \} ;$$

thus each W_n is open, the sets $\{W_n\}_{n\geq 1}$ are disjoint, and Lemma 7.2(1) shows that

(7.11) $\quad U_{m_n} \subset W_{\ell_n}$ for all $n \geq 0$,

where $\ell_n = m_{n+1}-m_n$.

Lemma 7.3 Let J be a (maximal connected) component of W_n . Then

(1) f^n is monotone on J .

(2) If z is an end-point of $\bar{J}$ and $z \in (0,1)$ then $f^{n+1}(z) = f(z)$.

(3) φ is not an end-point of $\bar{J}$.

Proof (1): Let $k < n$ and $y \in J$; then $f^{k+1}(y) < f(y) \leq f(\varphi)$ and so $f^k(y) \neq \varphi$. Thus by Proposition 2.1 f^n is monotone on J .

(2): Let z be an end-point of $\bar{J}$ with $z \in (0,1)$; then for some $1 \leq k \leq n$ we have $f^{j+1}(z) < f(z)$ for $1 \leq j < k$ and $f^{k+1}(z) = f(z)$. Suppose n is not a multiple of k , thus $n = k\ell+j$ with $\ell \geq 1$ and $1 \leq j < k$, and hence $f^{n+1}(z) = f^j(f^{k\ell}(f(z))) = f^{j+1}(z)$. But $f^{j+1}(z) < f(z)$ (since $j < k$) and $f^{n+1}(z) \geq f(z)$; therefore this case cannot occur and so $n = k\ell$ for some $\ell \geq 1$. This gives us $f^{n+1}(z) = f^{k\ell}(f(z)) = f(z)$.

(3): Suppose φ is an end-point of $\bar{J}$; then by (2) we have $f^{n+1}(\varphi) = f(\varphi)$ and this implies that $f^n(\varphi) = \varphi$. Without loss of generality we can assume φ is the right-hand end-point of $\bar{J}$, thus

$\overline{J} = [a,\varphi]$ with $0 \leq a < \varphi$. By (1) f^n is monotone on $[a,\varphi]$; suppose first that f^n is increasing on this interval. If $a > 0$ then from (2) we have $f^{n+1}(a) = f(a)$ and thus $f^n(a) = b$, where b is such that $f(a) = f(b)$. Hence we have $f^n(a) = a$ (because $f^n(a) < f^n(\varphi) = \varphi$ and a is the only point in $[0,\varphi)$ whose image is $f(a)$), and this means that $[a,\varphi]$ is a sink. If $a = 0$ then $[a,\varphi]$ is again a sink. But by Lemma 7.1 $[a,\varphi]$ cannot be a sink (since f would then be central); f^n must therefore be decreasing on $[a,\varphi]$. However, it is clear from the definition of W_n that W_n also has a component I with φ the left-hand end-point of $\overline{I}$. If f^n is decreasing on $[a,\varphi]$ then it is increasing on $\overline{I}$ and the above proof now shows that $\overline{I}$ is a sink. Again this is not possible and thus φ cannot be an end-point of $\overline{J}$. ▫▫

Remark: If J is a component of W_n and z is an end-point of $\overline{J}$ then by Lemma 7.3(2) either $z \in \{0,1\}$ or $f(z) \in Per(n,f)$. Hence $1 + |Per(n,f)|$ is an upper bound for the number of components of W_n , and from Proposition 4.4 we know that $|Per(n,f)|$ is finite. Thus W_n has only finitely many components for each $n \geq 1$.

Note that we must have $f(\varphi) > \varphi$ (since if $f(\varphi) \leq \varphi$ then f would be central, and by Lemma 7.1 this is not the case). As usual let β denote the unique fixed point of f in $(\varphi,1)$. Now choose $0 < \eta < \varphi$ so that $f(\eta) > \beta$. Thus for each $x \in [\eta,\varphi)$ there exists a unique $\underline{x} \in (\varphi,\beta)$ with $f(\underline{x}) = f(x)$. (See the picture at the top of the next page.) For $y \in [\varphi,\underline{\eta}]$ let $\underline{y}$ denote the unique point in $[\eta,\varphi]$ with $f(\underline{y}) = f(y)$. In the case when φ is periodic we will always assume that η has been chosen close enough to φ so that $[\varphi] \cap [\eta,\underline{\eta}] = \{\varphi\}$ (i.e. so that $f^k(\varphi) \in [\eta,\underline{\eta}]$ implies $f^k(\varphi) = \varphi$). *Note:* If φ is periodic then it must be an unstable periodic point (otherwise f would be central).

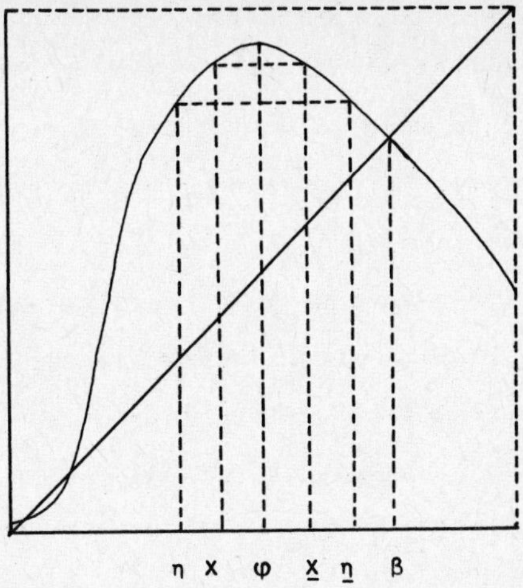

Lemma 7.4 Let J be a component of W_n and let z be an end-point of $\overline{J}$. If $z \in [\eta, \underline{\eta}]$ then z is not a turning point of f^n .

Proof Suppose z is a turning point of f^n ; by Proposition 2.1 and Lemma 7.3(3) there then exists $1 \leqq k < n$ with $f^k(z) = \varphi$. Now from Lemma 7.3(2) we have $f^{n+1}(z) = f(z)$ and so $f(z)$ is periodic. Thus φ is also periodic since $\varphi = f^{k-1}(f(z))$. But $f(f^n(z)) = f(z)$ and hence $f^n(z)$ is either z or $\underline{z}$, i.e. one of z and $\underline{z}$ lies in the orbit $[\varphi]$. However this is not possible because both of z and $\underline{z}$ are in $[\eta, \underline{\eta}]$. ⊞

We are assuming (7.2) and (7.3) hold; there thus exists $0 < \Delta \leqq 1$ with

(7.12) $\qquad \Delta f'(x) \leqq |f'(\underline{x})| \leqq (1/\Delta) f'(x) \quad$ for all $x \in [\eta, \varphi)$.

Lemma 7.5 Let J be a component of W_n and suppose $J \subset (\eta, \underline{\eta})$. Then $|(f^n)'(x)| \geqq \Delta$ for all $x \in J$.

Proof Let $J = (a,b)$ and note that $(\underline{b},\underline{a})$ is another component of W_n. By Lemma 7.3(2) we have that either $f^n(a) = a$ or $f^n(\underline{a}) = \underline{a}$; suppose first that $f^n(a) = a$. We have $a \geq \eta$ and thus a is not a fixed point of f in $[0,\varphi]$. Therefore Proposition 5.6 and Lemma 7.1 imply that a is an unstable fixed point of f^n . But by (7.6) and Lemma 7.4 f^n is continuously differentiable in a neighbourhood of a ; thus $|(f^n)'(a)| \geq 1$. If $f^n(\underline{a}) = \underline{a}$ then in the same way $|(f^n)'(\underline{a})| \geq 1$ and in this case we have $|(f^n)'(a)| \geq \Delta$ because

$$|(f^n)'(a)| = |\prod_{k=0}^{n-1} f'(f^k(a))|$$

$$= \left|\frac{f'(a)}{f'(\underline{a})}\right| \cdot |\prod_{k=0}^{n-1} f'(f^k(\tilde{a}))| = \left|\frac{f'(a)}{f'(\underline{a})}\right| \cdot |(f^n)'(\underline{a})| .$$

Thus in both cases we have $|(f^n)'(a)| \geq \Delta$ and replacing a with b we also get $|(f^n)'(b)| \geq \Delta$. Now let $[u,v]$ be the largest interval containing $[a,b]$ on which f^n is monotone, so by Lemma 7.4 we have $u < a < b < v$. Lemma 7.3(1) and Lemma 4.1 imply that the restriction of f^n to $[u,v]$ has property R and so by Propositions 4.6 and 4.8 we have $|(f^n)'(x)| \geq \min\{|(f^n)'(a)|,|(f^n)'(b)|\} \geq \Delta$ for all $x \in (a,b)$. ▫▫

For each $n \geq 0$ let J_n be the component of W_{ℓ_n} which contains U_{m_n} (recalling that (7.11) holds).

Lemma 7.6 There exists $N \geq 0$ such that $J_n \subset (\eta,\underline{\eta})$ for all $n \geq N$.

Proof We have $\lim_{n\to\infty} f^{m_n}(x) = \varphi$ for all $x \in U$ and there thus exists $k \geq 0$ such that $U_{m_k} \cap (\eta,\underline{\eta}) \neq \phi$. By Lemma 7.3(3) and the remark following Lemma 7.3 we can then find $\xi \in (\eta,\varphi)$ with $W_{\ell_k} \cap (\xi,\underline{\xi}) = \phi$. Now let

$N > k$ be such that $U_{m_N} \cap (\xi,\underline{\xi}) \neq \phi$. If $n > N$ then, since U_{m_n} is higher than U_{m_N}, we also have $U_{m_n} \cap (\xi,\underline{\xi}) \neq \phi$; therefore $J_n \cap (\xi,\underline{\xi}) \neq \phi$ for all $n \geqq N$. It thus easily follows that $J_n \subset (\eta,\underline{\eta})$ for all $n \geqq N$. (If $J_n \not\subset (\eta,\underline{\eta})$ for some $n \geqq N$ then either $(\eta,\xi] \subset J_n$ or $[\underline{\xi},\underline{\eta}) \subset J_n$, and hence $(\eta,\xi] \cup [\underline{\xi},\underline{\eta}) \subset W_{\ell_n}$ because $J_n \subset W_{\ell_n}$ and since if $x \in (\eta,\varphi)$ then $x \in W_{\ell_n}$ if and only if $\underline{x} \in W_{\ell_n}$. But this would imply that $W_{\ell_k} \cap W_{\ell_n} \neq \phi$ and thus that $\ell_n = \ell_k$. However, by construction we have $W_{\ell_k} \cap (\xi,\underline{\xi}) = \phi$ and $W_{\ell_n} \cap (\xi,\underline{\xi}) \neq \phi$.) ▫▫

Lemma 7.7 Let $N \geq 0$ be as in Lemma 7.6. Then for all $n \geq N$ we have $|U_{m_{n+1}}| \geqq \Delta |U_{m_n}|$.

Proof For any $n \geq 0$ we have $U_{m_{n+1}} = f^{\ell_n}(U_{m_n})$ and $U_{m_n} \subset J_n \subset W_{\ell_n}$. If $n \geqq N$ then $J_n \subset (\eta,\underline{\eta})$ and so by Lemma 7.5 $|(f^{\ell_n})'(x)| \geqq \Delta$ for all $x \in J_n$; in this case the mean value theorem implies that $|U_{m_{n+1}}| \geqq \Delta |U_{m_n}|$. ▫▫

At this point it is worth mentioning that Lemma 7.7 immediately gives us the following result:

Proposition 7.2 Suppose $f \in S_R$ satisfies (7.2) and

(7.13) there exists $\delta > 0$ such that $f(\varphi - x) = f(\varphi + x)$ for all $x \in (0,\delta)$.

Then f has no type 1 homtervals.

Proof Let $\delta > 0$ be such that (7.13) holds. If we choose η so that $\eta \in (\varphi-\delta,\varphi)$ then $f'(\underline{x}) = -f'(x)$ for all $x \in [\eta,\varphi)$, and hence (7.12) holds with $\Delta = 1$. By Lemma 7.7 we thus have $|U_{m_{n+1}}| \geqq |U_{m_n}|$ for all $n \geqq N$, and this gives us a contradiction because the intervals $\{U_{m_n}\}_{n \geqq N}$ are disjoint. ∎∎

Remark: The hypotheses of Proposition 7.2 are of course satisfied by the functions in the families $f_\mu(x) = \mu x(1-x)$, $0 < \mu \leq 4$, and $f_\mu(x) = \sin(\mu x)$, $\frac{\pi}{2} < \mu \leq \pi$.

We now continue with the proof of Proposition 5.9(2) in the general case (i.e. when we only have $\Delta < 1$). Put $U = (a,b)$ and for each $n \geqq 0$ let $K_n = [a_n, b_n]$ be the largest interval containing U on which f^n is monotone.

Lemma 7.8 (1) For each $n \geqq 0$ we have $a_n \leqq a_{n+1} < a < b < b_{n+1} \leqq b_n$.

(2) $a = \lim_{n\to\infty} a_n$ and $b = \lim_{n\to\infty} b_n$.

Proof (1): Clearly $a_n \leqq a_{n+1}$ and $b_{n+1} \leqq b_n$. By (7.9) we have $(a,b) \subset (0,1)$ and thus if we had either $a_m = a$ or $b_m = b$ for some $m \geqq 0$ then by Proposition 2.1 there would exist $0 \leqq k < m$ with $\varphi \in f^k(\overline{U}) = \overline{U}_k$; however, this contradicts (7.8).

(2): If $\overline{a} = \lim_{n\to\infty} a_n$ and $\overline{b} = \lim_{n\to\infty} b_n$ then $(\overline{a},\overline{b}) \subset H$; thus $(\overline{a},\overline{b}) = (a,b)$ since $(a,b) = U$ is a component of H . ∎∎

By Lemma 7.8(2) we can find $\overline{N} \geqq N$ (with N as in Lemma 7.6) such that $[a_{m_{\overline{N}}}, b_{m_{\overline{N}}}] \subset (0,1)$ and

(7.14) $|a_{m_n} - a| < \Delta^2(b-a)$ and $|b_{m_n} - b| < \Delta^2(b-a)$ for all $n \geqq \overline{N}$.

The rest of the proof will consist in showing that $|U_{m_{n+1}}| \geqq |U_{m_n}|$ for all $n \geqq N$.

For each $n \geqq 0$ let $L_n = [a_n, a]$ and $R_n = [b, b_n]$, thus $K_n = L_n \cup U \cup R_n$ and by Lemma 7.8(1) L_n and R_n are non-trivial intervals. If I is an interval with $I \subset [0,\varphi]$ (resp. $I \subset [\varphi,1]$) then let $\underline{I} = [\varphi,1] \cap f^{-1}(f(I))$ (resp. $\underline{I} = [0,\varphi] \cap f^{-1}(f(I))$); in particular, if $I = [c,d]$ is a sub-interval of either $[n,\varphi]$ or $[\varphi,\underline{n}]$ then $\underline{I} = [\underline{d},\underline{c}]$. Let $n_0 \geqq 0$ be such that $[a_{n_0}, b_{n_0}] \subset (0,1)$.

Lemma 7.9 $\varphi \in f^{m_n}(K_{m_n})$ for each $n \geqq n_0$.

Proof Let $n \geqq n_0$, put $p = m_n$ and suppose $\varphi \notin f^p(K_p)$. Thus either $f^p(R_p)$ or $f^p(L_p)$ lies between φ and $f^p(U) = U_p$; without loss of generality assume it is the former. By Proposition 2.1 there exists $0 \leqq j < p$ with $f^j(b_p) = \varphi$, and then the end-points of the interval $f^j(R_p)$ are φ and $f^j(b)$. Since U_p is higher than U_j this implies that either $U_p \subset f^j(R_p)$ or $U_p \subset \underline{f^j(R_p)}$. Now $f^p(R_p)$ lies between φ and U_p , and both of $f^j(R_p)$ and $\underline{f^j(R_p)}$ have φ as an end-point; thus either $f^p(R_p) \subset f^j(R_p)$ or $f^p(R_p) \subset \underline{f^j(R_p)}$. Therefore f^{p-j} maps either $f^j(R_p)$ or $\underline{f^j(R_p)}$ (monotonically) back into itself, and hence one of these intervals is a sink. But this would mean that f is central, which by Lemma 7.1 is not the case. ⊞

Note that $\varphi \notin U_{m_n} = f^{m_n}(U)$ and thus Lemma 7.9 says that either $\varphi \in f^{m_n}(L_{m_n})$ or $\varphi \in f^{m_n}(R_{m_n})$.

Lemma 7.10 Let $n \geqq n_0$ and put $p = m_n$, $q = m_{n+1}$.

(1) If $\varphi \in f^q(R_q)$ then either $U_p \subset f^q(L_q)$ or $\underline{U_p} \subset f^q(L_q)$.

(2) If $\varphi \in f^q(L_q)$ then either $U_p \subset f^q(R_q)$ or $\underset{\smile}{U_p} \subset f^q(R_q)$.

Proof (1): Assume that $\varphi \in f^q(R_q)$, so $\varphi \notin f^q(L_q)$. Let J be the closed interval which fills in the gap between U_p and φ (thus $U_p \cap J = \phi$ and the end-points of J are φ and either $f^p(a)$ or $f^p(b)$). Since U_q is higher than U_p we have either $U_q \subset J$ or $U_q \subset \underset{\smile}{J}$. Now by Proposition 2.1 there exists $0 \leq j < q$ with $f^j(a_q) = \varphi$; the end-points of $f^j(L_q)$ are then φ and $f^j(a)$. But $j < q$ and so either $j = p$ or U_p is higher than U_j , and this implies that either $J \subset f^j(L_q)$ or $\underset{\smile}{J} \subset f^j(L_q)$. Hence $f^{q-j}(J) \subset f^q(L_q)$. Suppose that $f^q(a_q) \in U_p \cup J$; then $U_q \subset J$ (since $U_q \subset \underset{\smile}{J}$ would imply that $\varphi \in f^q(L_q)$) and thus $f^q(L_q) \subset U_p \cup J$. Therefore

$$f^{q-j}(J) \subset f^{q-j}(f^j(L_q)) = f^q(L_q) \subset U_p \cup J ,$$

and this gives us $f^{q-j}(U_p \cup J) \subset U_p \cup J$. (Note that $f^{q-j}(U_p) = U_{q-j+p}$ and $\overline{f^{q-j}(U_p)}$ has at least one of its end-points in $U_p \cup J$; thus U_{q-j+p} is higher than U_p and so in fact $f^{q-j}(U_p) \subset J$.) However, f^{q-j} is monotone on $U_p \cup J$: f^{q-j} is clearly monotone on U_p and it is monotone on J because either J or $\underset{\smile}{J}$ is a subset of $f^j(L_q)$; also if an end-point of $\overline{U}_p$ was a turning point of f^{q-j} then this would contradict (7.8). Hence $\overline{U_p \cup J}$ is a sink. But this is not possible because it would imply f is central; therefore $f^q(a_q) \notin U_p \cup J$. Using the same argument we obtain $f^q(a_q) \notin \underset{\smile}{U_p \cup J}$. This, together with the fact that $f^q(a) \in J \cup \underset{\smile}{J}$ and $\varphi \notin f^q(L_q)$, shows that either $U_p \subset f^q(L_q)$ or $\underset{\smile}{U_p} \subset f^q(L_q)$.

(2): This is the same as (1). $\square\square$

The final ingredient in the proof of Proposition 5.9(2) is the following:

Lemma 7.11 For each $n \geqq 0$ we have

$$\frac{|f^n(U)|}{|U|} \geqq \min\left\{\frac{|f^n(L_n)|}{|L_n|}, \frac{|f^n(R_n)|}{|R_n|}\right\}.$$

Proof f^n is continuously differentiable in (a_n,b_n), so by the mean value theorem there exist $x \in (a_n,a)$ and $y \in (b,b_n)$ with $|(f^n)'(x)| = |f^n(L_n)|/|L_n|$ and $|(f^n)'(y)| = |f^n(R_n)|/|R_n|$. Now the restriction of f^n to $[a_n,b_n]$ has property R and hence by Propositions 4.6 and 4.8 we have $|(f^n)'(z)| \geqq \min\{|(f^n)'(x)|,|(f^n)'(y)|\}$ for all $z \in (a,b)$. This clearly gives us what we want. □□

Now let $n \geqq \overline{N}$ and put $p = m_n$, $q = m_{n+1}$ and $r = m_{n+2}$. By Lemma 7.9 we have either $\varphi \in f^q(L_q)$ or $\varphi \in f^q(R_q)$; without loss of generality let us assume that $\varphi \in f^q(R_q)$. Then (since U_r is higher than U_q) either $U_r \subset f^q(R_q)$ or $\underset{\smile}{U_r} \subset f^q(R_q)$, and hence

$$|f^q(R_q)| \geqq \min\{|U_r|,|\underset{\smile}{U_r}|\} \geqq \Delta|U_r|.$$

($|\underset{\smile}{U_r}| \geqq \Delta|U_r|$ follows from (7.12) because $U_r \subset (\eta,\underline{\eta})$.) But $r = m_{n+2}$ and $q = m_{n+1}$ and so Lemma 7.7 gives us that $|U_r| \geqq \Delta|U_q|$. Therefore $|f^q(R_q)| \geqq \Delta^2|U_q|$ and thus using (7.14) we have

$$\frac{|R_q|}{|U|} = \frac{(b_q-b)}{(b-a)} < \Delta^2 \leqq \frac{|f^q(R_q)|}{|U_q|} = \frac{|f^q(R_q)|}{|f^q(U)|}.$$

Hence $|f^q(U)|/|U| < |f^q(R_q)|/|R_q|$ and so Lemma 7.11 now implies that

$|f^q(U)|/|U| \geq |f^q(L_q)|/|L_q|$, and thus again using (7.14) we get

$$|U_q| = |f^q(U)| \geq (1/\Delta^2)|f^q(L_q)| \ .$$

Finally, we can apply Lemma 7.10: Since we are assuming $\varphi \in f^q(R_q)$ this result tells us that either $U_p \subset f^q(L_q)$ or $\underset{\smile}{U_p} \subset f^q(L_q)$, and thus

$$|f^q(L_q)| \geq \min\{|U_p|, |\underset{\smile}{U_p}|\} \geq \Delta|U_p| \ .$$

This, together with the previous inequality, gives us

$$|U_q| \geq (1/\Delta^2)(\Delta|U_p|) = (1/\Delta)|U_p| \geq |U_p| \ ;$$

i.e. $|U_{m_{n+1}}| \geq |U_{m_n}|$.

The proof of Proposition 5.9(2) is therefore now complete.

Notes: Proposition 5.9 is almost the same as Proposition 2.6 in Guckenheimer (1979), and the presentation given here is adapted from Guckenheimer's proof.

8. KNEADING SEQUENCES

Let $\mathcal{W} = \{ \{\theta_n\}_{n \geqq 0} : \theta_n \in \{-1,0,1\}$ for each $n \geqq 0 \}$ and for $f \in \mathcal{S}$ define $k_f = \{k_f(n)\}_{n \geqq 0} \in \mathcal{W}$ by

$$k_f(n) = \begin{cases} -1 & \text{if } f^n(\varphi) < \varphi \, , \\ 0 & \text{if } f^n(\varphi) = \varphi \, , \\ 1 & \text{if } f^n(\varphi) > \varphi \, . \end{cases}$$

k_f is called the *kneading sequence* of f and in this section we consider how much k_f tells us about a function $f \in \mathcal{S}$. The main result will show that if $f \in \mathcal{S}_R^*$ with $f(z) > z$ for all $z \in (0,\varphi]$ is such that φ is not periodic then we can determine which one of (5.6), (5.7) and (5.8) holds from just knowing k_f. (If φ is periodic then Proposition 5.4 gives us that (5.6) holds (resp. (5.7) holds) if and only if φ is stable (resp. φ is unstable), and the kneading sequence by itself does not provide enough information to distinguish between these two cases.)

Let $T : \mathcal{W} \longrightarrow \mathcal{W}$ be given by $T(\{\theta_n\}_{n \geqq 0}) = \{\theta_{n+1}\}_{n \geqq 0}$, and for $m \geqq 0$ define $T^m : \mathcal{W} \rightarrow \mathcal{W}$ inductively by $T^0(\theta) = \theta$, $T^1 = T$ and $T^m(\theta) = T(T^{m-1}(\theta))$; thus in fact we have $T^m(\{\theta_n\}_{n \geqq 0}) = \{\theta_{n+m}\}_{n \geqq 0}$. $\theta \in \mathcal{W}$ is said to be *periodic* if $T^m(\theta) = \theta$ for some $m \geqq 1$, and the smallest such $m \geqq 1$ is then called the *period* of θ. If $f \in \mathcal{S}$ then it is clear that φ is periodic with period n if and only if k_f is periodic with period n.

Remark: (8.1) Let $f \in \mathcal{S}_R^*$ with $f(z) > z$ for all $z \in (0,\varphi]$ and suppose that φ is not periodic. By Proposition 5.4 we have that (5.6) holds if and only if φ is regular, and it is easy to see that φ is regular if and only if $T(k_f)$ is periodic.

For $m > 1$ we define $D_m : W \longrightarrow W$ and $H_m : W \rightarrow W$ by

$$D_m(\{\theta_n\}_{n\geq 0}) = \{\mu_m^n\theta_n\}_{n\geq 0}\ , \text{ where } \mu_m^n = \begin{cases} 0 & \text{if } n = 0 \pmod m\ , \\ 1 & \text{otherwise,} \end{cases} \quad \text{and}$$

$H_m(\{\theta_n\}_{n\geq 0}) = \{\theta_{mn}\}_{n\geq 0}$. Let

$$W_1 = \{\ \theta \in W : D_m(\theta) \text{ is periodic with period } m \text{ for some } m > 1\ \}\ ,$$

and for $\theta \in W_1$ let $p(\theta)$ be the smallest integer $m > 1$ such that $D_m(\theta)$ is periodic with period m . We define $N : W_1 \longrightarrow W$ by $N(\theta) = H_{p(\theta)}(\theta)$ and for $m \geq 1$ let $W_m \subset W$ and $N^m : W_m \longrightarrow W$ be given inductively by $N^1 = N$, $W_m = \{\ \theta \in W_{m-1} : N^{m-1}(\theta) \in W_1\ \}$ and $N^m(\theta) = N(N^{m-1}(\theta))$; put $W_\infty = \bigcap_{m\geq 1} W_m$.

For $\varepsilon = -1, 1$ let $\Delta_\varepsilon = \{\Delta_\varepsilon(n)\}_{n\geq 0}$, where $\Delta_\varepsilon(0) = 0$ and $\Delta_\varepsilon(n) = \varepsilon$ for all $n \geq 1$; let $E = \{\Delta_{-1},\Delta_1\}$. Note that $E \subset W_1$ and $N(\theta) = \theta$ for each $\theta \in E$; thus $E \subset W_\infty$. Let

$$E_\infty = \{\ \theta \in W_\infty : N^m(\theta) \in E \text{ for some } m \geq 1\ \}\ .$$

Theorem 8.1 Let $f \in S_R^*$ with $f(z) > z$ for all $z \in (0,\varphi]$ and suppose that φ is not periodic. Then:

(1) (5.6) holds if and only if $k_f \in E_\infty$.

(2) (5.7) holds if and only if $k_f \notin W_\infty$.

(3) (5.8) holds if and only if $k_f \in W_\infty - E_\infty$.

Proof This will be a special case of Theorem 8.2. ▯▯

Remark: Let $\theta = \{\theta_n\}_{n\geq 0} \in W$ with $\theta_0 = 0$ and $\theta_n \neq 0$ for all $n \geq 1$. Then it is not difficult to check that $\theta \in E_\infty$ if and only if $T(\theta)$ is periodic. This fact gives us the connection between Theorem 8.1(1) and Remark (8.1).

Theorem 8.1 corresponds to Theorem 5.1; we now consider a more general result which corresponds to Theorem 5.2. Let $f \in S$; then clearly we have $f(\varphi) \leqq \varphi$ if and only if $k_f \in \{\Delta_{-1}, \Delta_0\}$, where $\Delta_0 = \{\Delta_0(n)\}_{n \geqq 0}$ is such that $\Delta_0(n) = 0$ for all $n \geqq 0$. If $f(\varphi) > \varphi$ then it is also clear that $f^2(\varphi) \leqq \gamma$ if and only if $k_f = \{\tau_n\}_{n \geqq 0}$, where $\tau_0 = 0$, $\tau_1 = 1$ and $\tau_n = -1$ for all $n \geqq 2$. (Note that we cannot distinguish between $f^2(\varphi) < \gamma$ and $f^2(\varphi) = \gamma$ from just knowing k_f .)

Theorem 8.2 Let $f \in S$ with $f(\varphi) > \varphi$ and $f^2(\varphi) \geqq \gamma$, and suppose that φ is not periodic. Then:

(1) (5.11) holds if and only if $k_f \in E_\infty$.

(2) Either (5.12) or (5.13) holds if and only if $k_f \notin W_\infty$.

(3) (5.14) holds if and only if $k_f \in W_\infty - E_\infty$.

Proof We will use the notation and definitions from Section 6. Let $f \in S$ with $f(\varphi) > \varphi$ and $f^2(\varphi) \geqq \gamma$; by Proposition 6.4 and Lemma 6.3 we have:

(8.1) (5.11) holds if and only if $V^n f \in S_1$ for some $n \geqq 0$.

(8.2) Either (5.12) or (5.13) holds if and only if $V^n f \in S_4 \cup S_5 \cup S_6$ for some $n \geqq 0$.

(8.3) (5.14) holds if and only if $V^n f \in S_2 \cup S_3$ for all $n \geqq 0$.

We must therefore show that if f satisfies the hypotheses of Theorem 8.2 then:

(8.4) $V^n f \in S_1$ for some $n \geqq 0$ if and only if $k_f \in E_\infty$.

(8.5) $V^n f \in S_4 \cup S_5 \cup S_6$ for some $n \geqq 0$ if and only if $k_f \notin W_\infty$.

(8.6) $V^n f \in S_2 \cup S_3$ for all $n \geqq 0$ if and only if $k_f \in W_\infty - E_\infty$.

(8.4), (8.5) and (8.6) will follow from the next result.

Proposition 8.1 Let $f \in S$ be such that φ is not periodic. Then:

(8.7) $f \in S_1$ if and only if $k_f = \Delta_{-1}$.

(8.8) If $f \in S_2$ then $k_f \in W_1$ and $p(k_f) = 2$; also $k_{Vf} = -N(k_f)$. (If $\theta = \{\theta_n\}_{n \geqq 0} \in W$ then $-\theta = \{-\theta_n\}_{n \geqq 0}$.)

(8.9) If $k_f \in W_1$ and $p(k_f) = 2$ then either *(i)* $f \in S_1 \cup S_2$, or *(ii)* $f \in S_4 \cup S_5$ and $-N(k_f) = \tau = \{\tau_n\}_{n \geqq 0}$, where $\tau_0 = 0$, $\tau_1 = 1$ and $\tau_n = -1$ for all $n \geqq 2$.

(8.10) $f \in S_3$ if and only if $k_f \in W_1$ and $p(k_f) > 2$; if $f \in S_3$ then k_{Vf} is either $N(k_f)$ or $-N(k_f)$.

Before proving Proposition 8.1 let us show how it implies Theorem 8.2. Let $f \in S$ be such that φ is not periodic; if $f \in S_2 \cup S_3$ then by (8.8) and (8.10) we have $k_f \in W_1$ and k_{Vf} is either $N(k_f)$ or $-N(k_f)$. Conversely, if $k_f \in W_1$ then by (8.9) and (8.10) either $f \in S_1 \cup S_2 \cup S_3$ or $f \in S_4 \cup S_5$ and $N(k_f) = -\tau$. But $-\tau \notin W_1$ and so if $N(k_f) = -\tau$ then $k_f \notin W_\infty$. Now if $\theta \in W_1$ then $-\theta \in W_1$ and $N(-\theta) = -N(\theta)$; also $-\Delta_{-1} = \Delta_1$. From this it easily follows that (8.4), (8.5) and (8.6) hold for any $f \in S$ satisfying the hypotheses of Theorem 8.2. (Note that if φ is not periodic then $\varphi(V^n f)$ is also not periodic for each $n \geqq 0$.) ☐☐

Proof of Proposition 8.1 (8.7) is clear. (We cannot have $f(\varphi) = \varphi$ because φ is not periodic.) We will prove (8.8), (8.9) and (8.10) with the help of a couple of lemmas.

Lemma 8.1 Let $f \in \mathcal{S}$ and $\Gamma = ([c,d],f^m)$ be a reduction of $([0,1],f)$. Then $D_m(k_f)$ is periodic with period m and

$$k_{U_\Gamma f} = \begin{cases} H_m(k_f) & \text{if } \prod_{n=1}^{m-1}(-k_f(n)) = 1 , \\ -H_m(k_f) & \text{otherwise.} \end{cases}$$

Proof For each $n = 1,2,\ldots,m-1$ we have either $f^n([c,d]) \subset [0,\varphi)$ or $f^n([c,d]) \subset (\varphi,1]$, and $f^{\ell m+n}(\varphi) \in f^n([c,d])$ for each $\ell \geqq 0$. Thus $D_m(k_f)$ is periodic with period m. Now we have $U_\Gamma f = \psi_\Gamma \circ g \circ \psi_\Gamma^{-1}$, where g is the restriction of f^m to $[c,d]$ and where $\psi_\Gamma : [c,d] \to [0,1]$ is increasing (resp. decreasing) if f^m is increasing (resp. decreasing) on $[c,\varphi]$. Thus

$$k_{U_\Gamma f} = \begin{cases} H_m(k_f) & \text{if } f^m \text{ is increasing on } [c,\varphi] , \\ -H_m(k_f) & \text{if } f^m \text{ is decreasing on } [c,\varphi] . \end{cases}$$

But it is easy to see that f^m is increasing on $[c,\varphi]$ if and only if $\prod_{n=1}^{m-1}(-k_f(n)) = 1$, and this gives us what we want. ▫▫

For $f \in \mathcal{S}$ and $x \in [0,1]$ we define $\theta_f(x) = \{\theta_f(x,n)\}_{n \geqq 0} \in \mathcal{W}$ by

$$\theta_f(x,n) = \begin{cases} -1 & \text{if } f^n(x) < \varphi , \\ 0 & \text{if } f^n(x) = \varphi , \\ 1 & \text{if } f^n(x) > \varphi ; \end{cases}$$

thus in particular we have $\theta_f(\varphi) = k_f$.

Lemma 8.2 Let $f \in \mathcal{S}$ be such that φ is not periodic and suppose that $D_m(k_f)$ is periodic with period $m > 1$. Let

$$J = \{ x \in [0,1] : D_m(\theta_f(x)) = D_m(k_f) \} .$$

Then J is a non-trivial interval with $f^m(J) \subset J$ and $\varphi \in \text{int}(J)$.

Proof We have

$$J = \{ x \in [0,1] : \theta_f(x,n) = k_f(n) \text{ for all } n \neq 0 \pmod m \},$$

and $k_f(n+m) = k_f(n)$ when $n \neq 0 \pmod m$ (because $D_m(k_f)$ has period m). Thus if $x \in J$ then for $n \neq 0 \pmod m$ we have

$$\theta_f(f^m(x),n) = \theta_f(x,n+m) = k_f(n+m) = k_f(n),$$

i.e. $f^m(x) \in J$, and so $f^m(J) \subset J$. Clearly we have $\varphi \in J$. Now for $x, y \in [0,1]$ let $]x,y[$ denote the closed interval with end-points x and y (i.e. $]x,y[$ is either $[x,y]$ or $[y,x]$). Let $x \in J$ and $z \in]\varphi,x[$; since $x, \varphi \in J$ we have $\varphi \notin \operatorname{int}(f^n(]\varphi,x[))$ for $n = 0,1,\ldots,m-1$, and so $f^n(z) \in]f^n(\varphi),f^n(x)[$ for $n = 1,2,\ldots,m$. Thus $\theta_f(z,n) = k_f(n)$ for $n = 1,2,\ldots,m-1$ (because we have $\theta_f(x,n) = k_f(n) = \theta_f(\varphi,n)$ for $n = 1,2,\ldots,m-1$). Since also $f^m(z) \in]f^m(\varphi),f^m(x)[$ we have $f^m(z) \in]\varphi,z'[$, where z' is either $f^m(\varphi)$ or $f^m(x)$. But $z' \in J$ and so exactly as above we obtain $\theta_f(z,n+m) = \theta_f(f^m(z),n) = k_f(n)$ for $n = 1,2,\ldots,m-1$ and $f^{2m}(z) \in]f^m(\varphi),f^m(z')[$. Iterating this procedure gives us $\theta_f(z,n+\ell m) = k_f(n)$ for $n = 1,2,\ldots,m-1$ and for all $\ell \geq 0$. Therefore $z \in J$ because $k_f(n+\ell m) = k_f(n)$ for $n = 1,2,\ldots,m-1$, and this shows that J is an interval. Finally, if $x, y \in [0,1]$ are such that $f(x) = f(y)$ then $x \in J$ if and only if $y \in J$; this implies that if $\varphi \notin \operatorname{int}(J)$ then $J = \{\varphi\}$. But $J = \{\varphi\}$ would imply that $f^m(\varphi) = \varphi$, which by assumption is not the case. ▫▫

We come now to the proofs of (8.8), (8.9) and (8.10):

(8.8): If $f \in \mathcal{S}_2$ then by Proposition 6.1 $\Gamma = ([\alpha,\beta],f^2)$ is a reduction of $([0,1],f)$, and so by Lemma 8.1 we have $D_2(k_f)$ is periodic with period 2 and $k_{U_\Gamma f} = -H_2(k_f)$. Thus $k_f \in \mathcal{W}_1$ and clearly $p(k_f) = 2$; hence $k_{Vf} = -N(k_f)$ (because $Vf = U_\Gamma f$).

(8.9): Suppose $k_f \in W_1$ with $p(k_f) = 2$ and $f \notin S_1 \cup S_2$. Since $f \notin S_1$ we have $k_f(1) = 1$ and thus $k_f(2n+1) = 1$ for all $n \geqq 0$. Let $J = \{ x \in [0,1] : \theta_f(x,2n+1) = 1$ for all $n \geqq 0 \}$; Lemma 8.2 gives us that J is an interval with $f^2(J) \subset J$, and since $\alpha, \beta \in J$ we have $[\alpha,\beta] \subset J$. Let ξ be the smallest fixed point of f^2 in $[\varphi,\beta]$; if $f^3(\varphi) < \xi$ then the proof of Lemma 6.15 would produce points in $[\alpha,\beta]$ which are not in J , and so $\xi \leqq f^3(\varphi)$. But $f \notin S_2$ and hence we have $\xi \leqq f^3(\varphi) < \beta$. It now easily follows that $f \in S_4 \cup S_5$ and $N(k_f) = -\tau$.

(8.10): If $f \in S_3$ then by Lemma 8.1 we have $k_f \in W_1$; also by (8.9) we must have $p(k_f) > 2$. Conversely, suppose that $k_f \in W_1$ with $p(k_f) > 2$; by (8.7) and (8.8) we then have $f \notin S_1 \cup S_2$. Let $\ell > 2$ be such that $D_\ell(k_f)$ is periodic with period ℓ and let $J = \{ x \in [0,1] : D_\ell(\theta_f(x)) = D_\ell(k_f) \}$; Lemma 8.2 gives us that J is an interval with $\varphi \in \operatorname{int}(J)$ and $f^\ell(J) \subset J$. Since $\ell > 2$ we have $\alpha \notin J$ and it thus follows that $J \subset D$, where (as in Section 6)

$$D = \{ x \in [0,1] : f^k(x) \in [0,\alpha) \text{ for infinitely many } k \geqq 0 \} .$$

Proposition 6.6 now shows that $f \in S_3$, and hence $f \in S_3$ if and only if $k_f \in W_1$ and $p(k_f) > 2$. Finally, let $f \in S_3$ and $\Gamma = ([c,d],f^m)$ be the simple reduction of smallest order satisfying (6.5). If $D_\ell(k_f)$ is periodic with period $\ell > 2$ then from the above proof and Proposition 6.6 we have that ℓ is a multiple of m . On the other hand Lemma 8.1 gives us that $D_m(k_f)$ is periodic with period m , and so $p(k_f) = m$. The second part of Lemma 8.1 now implies that k_{Vf} is either $N(k_f)$ or $-N(k_f)$. ◻◻

Notes: The kneading sequence of a function in S was introduced in

Milnor and Thurston (1977), although similar ideas also occur in Šarkovskii (1964), Metropolis, Stein and Stein (1973) and Štefan (1977). Theorem 8.1 is adapted from Guckenheimer (1979) and Jonker and Rand (1981). For much more concerning the theory of kneading sequences (in particular as regards the problem of topological conjugacy) the reader is recommended to look at the account in Collet and Eckmann (1980).

9. AN "ALMOST ALL" VERSION OF THEOREM 5.1

In this section we show that Theorem 5.1 more-or-less remains true if "contains a dense open subset of $[0,1]$ " is replaced by "has Lebesgue measure one". More precisely, we will prove the following result (where λ denotes Lebesgue measure defined on the Borel subsets of $[0,1]$):

Theorem 9.1 Let $f \in S_R^*$ with $f(z) > z$ for all $z \in (0,\varphi]$.

(1) If $|P_s(f)| = 1$ and $[x] \in P_s(f)$ is stable then $\lambda(A([x],f)) = 1$.

(2) If (5.8) holds and $(\{I^{(n)}\}_{n\geq 1}, f)$ is the corresponding proper infinite register shift then for each $n \geq 1$

$$\lambda(\{\, x \in [0,1] : f^k(x) \in \text{int}(I^{(n)}) \text{ for some } k \geq 0 \,\}) = 1 .$$

If $|P_s(f)| = 1$ and $[x] \in P_s(f)$ is one-sided stable then it is probably also true that $\lambda(A([x],f)) = 1$; however for this case we can only give a somewhat weaker result, viz:

Theorem 9.2 Let $f \in S_R^*$ with $f(z) > z$ for all $z \in (0,\varphi]$; suppose that $|P_s(f)| = 1$ and that $[x] \in P_s(f)$ is one-sided stable. Then $\lambda(A^*([x],f)) = 1$, where for a periodic point z we define

$$A^*([z],f) = \{\, y \in [0,1] : \text{for each } \varepsilon > 0 \text{ there exists } k \geq 0 \text{ with } |f^k(y)-z| < \varepsilon \,\} .$$

Remark: If z is a periodic point of f then $A([z],f) \subset A^*([z],f)$, and $A([z],f) = A^*([z],f)$ when z is stable. However, if z is not stable then it is possible that $A^*([z],f) - A([z],f)$ is uncountable.

We now begin the proof of Theorem 9.1; we will use the notation and definitions from Section 6. First let us consider $f \in S_R^*$

with $f \in S_3$ and let $\Gamma = ([c,d],f^m)$ be the simple reduction of smallest order satisfying (6.5). Since Γ is simple we have $|(f^m)'(e_\Gamma)| \geqq 1$; we call f *proper* if $|(f^m)'(e_\Gamma)| > 1$.

Lemma 9.1 If f is not proper then $[e_\Gamma]$ is one-sided stable.

Proof $[e_\Gamma]$ cannot be stable because Γ is simple. However, if $g \in S_R^*$ and $x \in Per(n,g) \cap (0,1)$ is such that $|(g^n)'(x)| = 1$ then x cannot be unstable. (Let $[a,b]$ be the largest interval containing x on which g^{2n} is increasing, and let h be the restriction of g^{2n} to $[a,b]$. By Proposition 4.8 h' has no local minimum in (a,b) and thus x is not an unstable fixed point of h ; hence x is not an unstable periodic point of g .) Thus $[e_\Gamma]$ must be one-sided stable. ⊞

The key to the proof of Theorem 9.1 is the following result concerning the reductions which were introduced in Section 6. Recall that if $\Gamma = ([c,d],f^m)$ is a reduction of $([0,1],f)$ then

$$C_\Gamma = \{ x \in [0,1] : f^n(x) \in (c,d) \text{ for some } n \geq 0 \} .$$

Theorem 9.3 Let $f \in S_R^*$ with $f(z) > z$ for all $z \in (0,\varphi]$.

(1) If $f \in S_2$ and $\Gamma = ([\alpha,\beta],f^2)$ is the reduction given in Proposition 6.1 then $\lambda(C_\Gamma) = 1$.

(2) If $f \in S_3$ is proper and $\Gamma = ([c,d],f^m)$ is the simple reduction of smallest order satisfying (6.5) then $\lambda(C_\Gamma) = 1$.

Proof Later. ⊞

Proof of Theorem 9.1(1) Let

$$G = \{ f \in S : |P_s(f)| = 1 \text{ and if } [x] \in P_s(f) \text{ then } \lambda(A([x],f)) = 1 \} .$$

We need a couple of lemmas.

Lemma 9.2 Let $f \in S_R^*$ with $f(z) > z$ for all $z \in (0,\varphi]$ and suppose that $Vf \in G$. Suppose also that if $f \in S_3$ then f is proper. Then $f \in G$.

Proof If $f \notin S_2 \cup S_3$ then $Vf = f$, and so we can assume that $f \in S_2 \cup S_3$. We thus have $Vf = U_\Gamma f$, where $\Gamma = ([c,d], f^m)$ is either the reduction given in Proposition 6.1 (if $f \in S_2$) or the simple reduction of smallest order satisfying (6.5) (if $f \in S_3$). As in Section 6 let $\Psi_\Gamma : [c,d] \longrightarrow [0,1]$ be given by

$$\Psi_\Gamma(x) = \begin{cases} \dfrac{x-c}{d-c} & \text{if } f^m \text{ is increasing on } [c,\varphi], \\ \dfrac{d-x}{d-c} & \text{if } f^m \text{ is increasing on } [\varphi,d]. \end{cases}$$

Let $[x] \in P_s(Vf)$ and put $F = \Psi_\Gamma^{-1}(A([x],Vf))$; then $F \subset [c,d]$ and since $Vf \in G$ we have $\lambda(F) = d-c$. Also put $z = \Psi_\Gamma^{-1}(x)$; it is clear that $[z] \in P_s(f)$ and

$$A([z],f) \supset \{ y \in [0,1] : f^n(y) \in F \text{ for some } n \geqq 0 \}.$$

Now we can write $C_\Gamma = \bigcup_{u \in U} C_u$ as a disjoint union of non-empty open intervals, and it is easy to see that for each $u \in U$ there exists $n(u) \geqq 0$ such that $f^{n(u)}$ maps C_u monotonically into (c,d). Since $f^{n(u)}$ has a continuous derivative in C_u we have that $\lambda(\{ y \in C_u : f^{n(u)}(y) \in F \}) = \lambda(C_u)$, and it thus follows that

$$\lambda(A([z],f)) \geqq \lambda(\{ y \in [0,1] : f^n(y) \in F \text{ for some } n \geqq 0 \})$$

$$\geqq \sum_{u \in U} \lambda(\{ y \in C_u : f^{n(u)}(y) \in F \}) = \sum_{u \in U} \lambda(C_u) = \lambda(C_\Gamma).$$

But Theorem 9.3 gives us that $\lambda(C_\Gamma) = 1$, and hence $f \in G$. ⊞

Let $S_R^{**} = \{ f \in S_R^* : f(z) > z \text{ for all } z \in (0,\varphi] \}$.

Lemma 9.3 (1) If $f \in S_R^*$ then $Vf \in S_R^*$.

(2) If $f \in S_R^{**}$ then $Vf \in S_R^{**} \cup S_1$.

(3) If $f \in S_R^{**}$ and $Vf \in S_1$ then $Vf \in G$.

Proof (1): This is clear.

(2): If $f \notin S_2 \cup S_3$ then $Vf = f$, and so we can assume that $f \in S_2 \cup S_3$. By (1) we need to show that if $Vf \notin S_1$ then $(Vf)(z) > z$ for all $z \in (0,\varphi(Vf)]$. For $f \in S_2$ this follows from Proposition 6.1(2) and Proposition 5.5; for $f \in S_3$ it is a consequence of Proposition 4.6 and the fact that the reduction involved in the definition of Vf is simple.

(3): Let $f \in S_R^{**}$ with $Vf \in S_1$; then Vf has at most one fixed point x in the interval $(0,\varphi(Vf)]$, and if x exists then $(Vf)(z) > z$ for all $z \in (0,x)$ and $(Vf)(z) < z$ for all $z \in (x,\varphi(Vf)]$. (If $f \in S_3$ then this follows, as in (2), from Proposition 4.6 and because the reduction involved is simple. If $f \in S_2$ then this is a consequence of Proposition 4.6 and the fact that if $y \in [\varphi,\beta]$ is a fixed point of f^2 then $f(y) \in [\beta,f(\varphi)]$ is also a fixed point (cf. the proof of Proposition 6.1(2)).) Thus if x exists then $A([x],Vf) \supset (0,1)$ and hence $Vf \in G$. If Vf has no fixed point in $(0,\varphi(Vf)]$ then $(Vf)(z) < z$ for all $z \in (0,\varphi(Vf)]$ and hence $A([0],Vf) = [0,1]$; again we have $Vf \in G$. ⊞

Now let $f \in S_R^{**}$ with $|P_s(f)| = 1$ and suppose that the element in $P_s(f)$ is stable; from the proof of Theorem 5.2 we know then that $V^n f \in S_1$ for some $n \geq 1$. Let $\ell = \min\{ n \geq 1 : V^n f \in S_1 \}$, thus $V^n f \in S_2 \cup S_3$ for $0 \leq n < \ell$. Hence by Lemma 9.3(2) we have that $V^n f \in S_R^{**}$ for all $0 \leq n < \ell$ and by Lemma 9.3(3) we have $V^\ell f \in G$. Also if $V^n f \in S_3$ for some $0 \leq n < \ell$ then Lemma 9.1 gives us that $V^n f$ is proper. Lemma 9.2 therefore implies that $V^n f \in G$ for all

$0 \leqq n < \ell$ and so in particular $f \in \mathcal{G}$. This gives us Theorem 9.1(1). ▫▫

Proof of Theorem 9.1(2) Let $f \in S_R^{**}$ and suppose that (5.8) holds. From the proof of Theorem 5.2 we have that $V^n f \in S_2 \cup S_3$ for all $n \geqq 0$ and so Lemma 9.3(2) implies that $V^n f \in S_R^{**}$ for all $n \geqq 0$. Moreover, if $V^n f \in S_3$ for some $n \geqq 0$ then Lemma 9.1 tells us that $V^n f$ is proper (since $|P_s(f)| = 0$). Now for each $n \geqq 1$ there exists a reduction $\Gamma_n = ([c_n,d_n],(V^{n-1}f)^{m_n})$ of $([0,1],V^{n-1}f)$, and $V^n f$ is a linear rescaling of the restriction of $(V^{n-1}f)^{m_n}$ to $[c_n,d_n]$. As in the proof of Proposition 6.4(5) let us rewrite this without the rescaling: define $\{a_n\}_{n\geqq 0}$, $\{b_n\}_{n\geqq 0}$ by $a_0 = 0$, $b_0 = 1$ and $a_{n+1} = (b_n - a_n)c_{n+1} + a_n$, $b_{n+1} = (b_n - a_n)d_{n+1} + a_n$; also let $q_n = \prod_{k=1}^{n} m_k$. We thus have $a_n \leqq a_{n+1} < \varphi < b_{n+1} \leqq b_n$, and the restriction of f^{q_n} to $[a_n,b_n]$ is a scaled down (or scaled down and turned upside down) version of $V^n f$. Let g_n denote the restriction of f^{q_n} to $[a_n,b_n]$ and let

$$[u_n,v_n] = \begin{cases} [g_n^2(\varphi),g_n(\varphi)] & \text{if } g_n \text{ is increasing on } [a_n,\varphi] , \\ [g_n(\varphi),g_n^2(\varphi)] & \text{if } g_n \text{ is increasing on } [\varphi,b_n] . \end{cases}$$

From the definition of $I^{(n)}$ given in the proof of Proposition 6.4(5) we have $[u_n,v_n] \subset I^{(n)}$ and thus

$$\{ x \in [0,1] : f^k(x) \in \operatorname{int}(I^{(n)}) \text{ for some } k \geqq 0 \}$$
$$\supset \{ x \in [0,1] : f^k(x) \in (u_n,v_n) \text{ for some } k \geqq 0 \} .$$

But $$\{ x \in [0,1] : f^k(x) \in (u_n,v_n) \text{ for some } k \geqq 0 \}$$
$$= \{ x \in [0,1] : f^k(x) \in (a_n,b_n) \text{ for some } k \geqq 0 \} ,$$

since $(V^n f)(z) > z$ for all $z \in (0,\varphi(V^n f)]$ implies that either

$g_n(y) > y$ for all $y \in (a_n,\varphi]$ (if g_n is increasing on $[a_n,\varphi]$) or $g_n(y) < y$ for all $y \in [\varphi,b_n)$ (if g_n is increasing on $[\varphi,b_n]$). It therefore suffices to show that

$$\lambda(\{ x \in [0,1] : f^k(x) \in (a_n,b_n) \text{ for some } k \geqq 0 \}) = 1$$

for each $n \geqq 1$. Fix $n \geqq 1$ and for $\ell = n-1,\ldots,1,0$ define $G_\ell \subset [a_\ell,b_\ell]$ by

$$G_{n-1} = \{ x \in [a_{n-1},b_{n-1}] : g^k_{n-1}(x) \in (a_n,b_n) \text{ for some } k \geqq 0 \}$$

and (for $\ell < n-1$)

$$G_{\ell-1} = \{ x \in [a_{\ell-1},b_{\ell-1}] : g^k_{\ell-1}(x) \in G_\ell \text{ for some } k \geqq 0 \} .$$

Clearly we have $G_0 \subset \{ x \in [0,1] : f^k(x) \in (a_n,b_n) \text{ for some } k \geqq 0 \}$ and so it is enough to prove that $\lambda(G_0) = 1$. However, this is a consequence of Theorem 9.3: As in the proof of Lemma 9.2 it is straightforward to show that $\lambda(G_{n-1}) = b_{n-1}-a_{n-1}$ and that (for $\ell < n-1$) $\lambda(G_{\ell-1}) = b_{\ell-1}-a_{\ell-1}$ follows from having $\lambda(G_\ell) = b_\ell-a_\ell$; thus in particular $\lambda(G_0) = b_0-a_0 = 1$. ◻◻

Proof of Theorem 9.3 (1): Let $f \in S_R^{**}$ with $f \in S_2$ and let $\Gamma = ([\alpha,\beta],f^2)$ be the reduction of $([0,1],f)$ given in Proposition 6.1. Since $f(z) > z$ for all $z \in (0,\varphi]$ it follows from Proposition 6.1 that

$$[0,1]-C_\Gamma \subset \{0,1\} \cup \{ z \in [0,1] : f^n(z) = \beta \text{ for some } n \geqq 0 \} .$$

Thus $[0,1]-C_\Gamma$ is countable and so $\lambda(C_\Gamma) = 1$.

(2): Let $f \in S_R^{**} \cap S_3$ be proper and let $\Gamma = ([c,d],f^m)$ be the simple reduction of smallest order satisfying (6.5). For $n \geqq 1$ put

$$E_n = \{ x \in [0,1] : f^k(x) \notin (c,d) \text{ for } k = 0,1,\ldots,n-1 \} ;$$

thus $\{E_n\}_{n\geqq 1}$ is a decreasing sequence of closed subsets of $[0,1]$ and

$\bigcap_{n\geq 1} E_n = [0,1]-C_\Gamma$. Put $E_\infty = \bigcap_{n\geqq 1} E_n$; we therefore need to show that $\lambda(E_\infty) = 0$. It is easy to see that each E_n is a finite union of closed intervals; let τ_n be the length of the largest interval contained in E_n .

Lemma 9.4 $\lim_{n\to\infty} \tau_n = 0$.

Proof Since $f \in S_R^{**}$ we have $\mathrm{Init}(f) = \mathrm{Trap}(f) = \mathrm{Homt}_2(f) = \phi$, and thus by (6.5)

$$\mathrm{int}([0,1]-C_\Gamma) = B_\Gamma \subset \mathrm{Init}(f)\cup\mathrm{Trap}(f)\cup\mathrm{Homt}_2(f) = \phi .$$

Therefore E_∞ is nowhere dense, and so $\lim_{n\to\infty} \tau_n = 0$. ◻◻

For $x \in E_n$ let $E_n(x)$ denote the component of E_n containing x . Choose $\xi \in (0,c)$ and $\eta \in (d,1)$ so that $\eta > \max\{f(c),f(d)\}$ and $\xi \leq f(\eta)$; put $\tilde{E}_n = \{ x \in E_n : E_n(x)\cap[\xi,\eta] \neq \phi \}$, thus $\bigcap_{n>1} \tilde{E}_n = E_\infty\cap[\xi,\eta]$.

Proposition 9.1 For each $N_o \geq 1$ there exists $N \geqq N_o$ such that $|(f^N)'(x)| \geqq 2$ for all $x \in \tilde{E}_N$.

Proof We proceed via a couple of lemmas.

Lemma 9.5 If $n \geqq 1$ and $x \in \partial E_n - \{0,1\}$ then $\lim_{k\to\infty} |(f^k)'(x)| = +\infty$.

Proof Let $x \in \partial E_n - \{0,1\}$; then clearly there exists $p \geq 0$ such that $f^p(x) \in \{c,d\}$ and so there exists $q \geqq 0$ with $f^q(x) = e_\Gamma$ (because $f^m(c) = f^m(d) = e_\Gamma$). Let $\ell \geqq 0$, $0 \leq i < m$; since $f^m(e_\Gamma) = e_\Gamma$ we have

$$|(f^{q+\ell m+i})'(x)| = |\prod_{t=0}^{q+\ell m+i-1} f'(f^t(x))|$$

$$= \left| \prod_{t=0}^{q-1} f'(f^t(x)) \left[\prod_{s=0}^{m-1} f'(f^s(e_\Gamma)) \right]^\ell \prod_{u=0}^{i-1} f'(f^u(e_\Gamma)) \right|$$

$$= \left| \prod_{t=0}^{q-1} f'(f^t(x)) \right| \left| \prod_{u=0}^{i-1} f'(f^u(e_\Gamma)) \right| \left| (f^m)'(e_\Gamma) \right|^\ell$$

$$\geqq \left| \prod_{t=0}^{q-1} f'(f^t(x)) \right| w^m \left| (f^m)'(e_\Gamma) \right|^\ell ,$$

where $w = \min_{0 \leqq u < m} |f'(f^u(e_\Gamma))|$. Therefore $\lim_{k \to \infty} |(f^k)'(x)| = +\infty$ because we are assuming that $|(f^m)'(e_\Gamma)| > 1$. □□

Note that $|(f^m)'(e_\Gamma)| > 1$ implies there exists $\gamma > 0$ such that $|f^{\ell m}([c,d])| \geqq \gamma$ for all $\ell \geqq 0$; there thus exists $\delta > 0$ such that $|f^k([c,d])| \geqq \delta$ for all $k \geqq 0$. By Lemma 9.4 we can choose $n_0 \geqq N_0$ so that $\tau_{n_0} \leqq \frac{1}{2} \min\{\delta, \xi, 1-\eta\}$, and by Lemma 9.5 there then exists $N \geqq n_0$ such that $|(f^N)'(x)| \geqq 2$ for all $x \in \partial E_{n_0} - \{0,1\}$. Now let $J = [u,v]$ be a component of $\tilde{E}_N$. By Proposition 2.1 J contains no turning point of f^N , so let $[a,b]$ be the maximal interval containing J on which f^N is monotone.

Lemma 9.6 There exist $y \in [a,u]$ and $z \in [v,b]$ with $|(f^N)'(y)| \geqq 2$ and $|(f^N)'(z)| \geqq 2$.

Proof Let $K = [\bar{u},\bar{v}]$ be the component of E_{n_0} containing J ; since $K \cap [\xi,\eta] \supset J \cap [\xi,\eta] \neq \phi$ and $|K| \leqq \tau_{n_0} \leqq \frac{1}{2} \min\{\xi, 1-\eta\}$ we have $\bar{u} > 0$ and $\bar{v} < 1$. We consider two cases:

(i) $\bar{u} \geqq a$: Here we have $\bar{u} \in [a,u]$, and $|(f^N)'(\bar{u})| \geqq 2$ because $\bar{u} \in \partial E_{n_0} - \{0,1\}$.

(ii) $\bar{u} < a$: Here we have $a \in (0,1)$ and so by Proposition 2.1 there exists $k < N$ with $f^k(a) = \varphi$. Now f^k is monotone on $[a,b]$ and

$f^k(J)\cap(c,d) = \phi$; there thus exists $\bar{a} \in (a,u]$ such that $f^k(\bar{a}) \in \{c,d\}$. We then have

$$|f^N([a,\bar{a}])| = |f^{N-k}(f^k([a,\bar{a}]))| = |f^{N-k}([c,d])| \geqq \delta ,$$

since $f^k([a,\bar{a}])$ is either $[c,\varphi]$ or $[\varphi,d]$ and $f([c,\varphi]) = f([\varphi,d]) = f([c,d])$. (Note that $f(c) = f(d)$ because $f^m(c) = f^m(d)$ and for each $1 \leqq k < m$ φ does not lie between $f^k(c)$ and $f^k(d)$.) But $|[a,\bar{a}]| \leqq |K| \leqq \tau_{n_0} \leqq \delta/2$, and so by the mean value theorem there exists $y \in (a,\bar{a}) \subset [a,u]$ with $|(f^N)'(y)| \geqq 2$.
Thus in both cases there exists $y \in [a,u]$ with $|(f^N)'(y)| \geqq 2$, and in the same way there exists $z \in [v,b]$ with $|(f^N)'(z)| \geqq 2$. ▫▫

Let g denote the restriction of f^N to $[a,b]$; thus g is strictly monotone, has property R and is continuously differentiable in (a,b) . Therefore by Proposition 4.8 we have $|g'(x)| \geqq 2$ for all $x \in [y,z]$, and hence in particular $|(f^N)'(x)| \geqq 2$ for all $x \in J$. This completes the proof of Proposition 9.1. ▫▫

The next result gives the main estimate needed to show that $\lambda(E_\infty) = 0$.

Proposition 9.2 There exist $N \geq 1$ and $\Lambda > 0$ such that if $n \geq N$ and J is a component of $\tilde{E}_n$ then $|(f^n)'(x)/(f^n)'(y)| \leqq \Lambda$ for all $x, y \in J$.

Proof Let $N \geqq 1$ be as in the proof of Proposition 9.1 ; thus $|(f^N)'(x)| \geqq 2$ for all $x \in \tilde{E}_N$ and $\tau_N \leqq \frac{1}{2}\min\{\xi,1-\eta\}$. Put $\bar{\eta} = \frac{1}{2}(1+\eta)$ and $\bar{\xi} = \min\{\xi/2,f(\bar{\eta})\}$, so $0 < \bar{\xi} < c$ and $d < \bar{\eta} < 1$. Put

$$\Omega = \max\{\ |f''(x)/f'(x)|\ :\ x \in [\bar{\xi},c]\cup[d,\bar{\eta}]\ \} ;$$

thus if x and y are both in the same component of $[\bar{\xi},c]\cup[d,\bar{\eta}]$ then by the mean value theorem $\left| \log|f'(x)| - \log|f'(y)| \right| \leq \Omega|x-y|$. We will show that Proposition 9.2 holds with $\Lambda = \exp(2N\Omega)$.

Lemma 9.7 Let $n \geq N$ and $x \in \tilde{E}_n$. Then:

(1) $f^k(x) \in [\bar{\xi},c]\cup[d,\bar{\eta}]$ for all $0 \leq k < n$.

(2) $f^j(x) \in \tilde{E}_N$ for all $0 \leq j \leq n-N$.

Proof (1): If $f(y) \in (\bar{\eta},1]$ then $y \in (c,d)$; if $f(y) \in [0,\bar{\xi})$ then $y \in [0,\bar{\xi})\cup(\bar{\eta},1]$. Thus if $y \in [\bar{\xi},c]\cup[d,\bar{\eta}]$ then $f(y) \in [\bar{\xi},\bar{\eta}]$. But $x \in [\bar{\xi},c]\cup[d,\bar{\eta}]$ because $J\cap[\xi,\eta] \neq \phi$ and $|J| \leq \tau_N \leq \min\{\xi-\bar{\xi},\bar{\eta}-\eta\}$; by induction we therefore have $f^k(x) \in [\bar{\xi},c]\cup[d,\bar{\eta}]$ for all $0 \leq k < n$.

(2): Exactly as in (1) we have that if $y \in [\xi,c]\cup[d,\eta]$ then $f(y) \in [\xi,\eta]$; thus if $z \in E_n\cap[\xi,\eta]$ then $f^k(z) \in [\xi,\eta]$ for all $0 \leq k < n$. Now put $J = E_n(x)$; since $x \in \tilde{E}_n$ we have $J\cap[\xi,\eta] \neq \phi$, and hence $f^k(J)\cap[\xi,\eta] \neq \phi$ for all $0 \leq k < n$. But clearly we have $f^j(E_n) \subset E_N$ for $0 \leq j \leq n-N$, and therefore $f^j(x) \in f^j(J) \subset \tilde{E}_N$ for all $0 \leq j \leq n-N$. ◻◻

Lemma 9.8 Let $n \geq N$ and J be a component of $\tilde{E}_n$. Then

$$\left|\frac{(f^n)'(x)}{(f^n)'(y)}\right| \leq \exp\left(\Omega \sum_{k=0}^{n-1} |f^k(x)-f^k(y)|\right) \quad \text{for all } x, y \in J .$$

Proof Let $x, y \in J$ and $0 \leq k < n$; by Lemma 9.7(1) we have $f^k(x), f^k(y) \in [\bar{\xi},c]\cup[d,\bar{\eta}]$, and in fact $f^k(x)$ and $f^k(y)$ must both be in the same component of $[\bar{\xi},c]\cup[d,\bar{\eta}]$ because $f^k(J)\cap(c,d) = \phi$. Thus $\left|\log|f'(f^k(x))| - \log|f'(f^k(y))|\right| \leq \Omega|f^k(x)-f^k(y)|$ and hence

$$\left|\frac{(f^n)'(x)}{(f^n)'(y)}\right| = \left|\prod_{k=0}^{n-1} \frac{f'(f^k(x))}{f'(f^k(y))}\right|$$

$$= \exp\left(\sum_{k=0}^{n-1} (\log|f'(f^k(x))| - \log|f'(f^k(y))|) \right)$$

$$\leq \exp\left(\Omega \sum_{k=0}^{n-1} |f^k(x)-f^k(y)| \right) . \qquad \square\square$$

Lemma 9.9 Let $n \geq N$ and J be a component of $\tilde{E}_n$. Then

$$\sum_{k=0}^{n-1} |f^k(x)-f^k(y)| \leq 2N \quad \text{for all } x, y \in J .$$

Proof Let $x, y \in J$ and $0 \leq j \leq n-N$; by Lemma 9.7(2) we have $f^j(x), f^j(y) \in \tilde{E}_N$, and in fact $f^j(x)$ and $f^j(y)$ must both lie in the same component of $\tilde{E}_N$ (since $f^k(f^j(J))\cap(c,d) = \phi$ for all $0 \leq k < N$). Thus by Proposition 9.1 and the mean value theorem we have

$$|f^{N+j}(x)-f^{N+j}(y)| = |f^N(f^j(x))-f^N(f^j(y))|$$

$$\geq 2|f^j(x)-f^j(y)| .$$

Iterating this we obtain that if $N\ell+j \leq n$ then

$$|f^{N\ell+j}(x)-f^{N\ell+j}(y)| \geq 2^\ell|f^j(x)-f^j(y)|$$

and hence that $|f^j(x)-f^j(y)| \leq 2^{-\ell}$. Therefore

$$\sum_{k=0}^{n-1} |f^k(x)-f^k(y)| = \sum_{k=n-N}^{n-1} |f^k(x)-f^k(y)| + \sum_{k=n-2N}^{n-1-N} |f^k(x)-f^k(y)| + \cdots$$

$$\leq N + \tfrac{1}{2}N + \tfrac{1}{4}N + \cdots = 2N . \qquad \square\square$$

Proposition 9.2 (with $\Lambda = \exp(2N\Omega)$) now follows immediately from Lemmas 9.8 and 9.9. $\square\square$

Let $J = [u,v]$ be a component of E_n ; we call J *regular* if $\min\{f^n(u),f^n(v)\} \leq c$ and $\max\{f^n(u),f^n(v)\} \geq d$. If J is regular then (since f^n is monotone on J) there exist unique $\overline{u}, \overline{v}$ with

$u \leqq \bar{u} < \bar{v} \leqq v$ such that $f^n([\bar{u},\bar{v}]) = [c,d]$; we then have that $J \cap E_{n+1}$ consists of the two intervals $[u,\bar{u}]$ and $[\bar{v},v]$. If J is regular then we put $\hat{J} = (\bar{u},\bar{v})$, i.e. $\hat{J} = J - E_{n+1}$. If J is not regular then (again since f^n is monotone on J) we have that either $J \cap E_{n+1} = \phi$ or $J \cap E_{n+1}$ consists of a single interval (having either u or v as one of its end-points). We call $x \in E_\infty$ *regular* if $E_n(x)$ is a regular component of E_n for infinitely many $n \geqq 1$; let us put $E_\infty^* = \{ x \in E_\infty : x$ is regular $\}$.

Lemma 9.10 $E_\infty - E_\infty^*$ is countable, and so in particular $\lambda(E_\infty - E_\infty^*) = 0$.

Proof To each $x \in E_\infty$ we assign a "code" $\sigma(x) = \{\sigma_n(x)\}_{n \geqq 1}$ by letting

$$\sigma_n(x) = \begin{cases} -1 & \text{if } E_{n-1}(x) \text{ is regular and } E_n(x) \text{ is the left-hand component of } E_{n-1}(x) \cap E_n , \\ 1 & \text{if } E_{n-1}(x) \text{ is regular and } E_n(x) \text{ is the right-hand component of } E_{n-1}(x) \cap E_n , \\ 0 & \text{if } E_{n-1}(x) \text{ is not regular.} \end{cases}$$

(where $E_0(x) = [0,1]$ and $[0,1]$ is defined to be regular). If $x, y \in E_\infty$ with $x \leqq y$ and $\sigma(x) = \sigma(y)$ then it is easy to see that $[x,y] \subset E_\infty$; hence by Lemma 9.4 we must have $x = y$, and thus different points of E_∞ have different codes. But if $x \in E_\infty - E_\infty^*$ then $\{ n \geqq 1 : \sigma_n(x) \neq 0 \}$ is finite, and therefore $E_\infty - E_\infty^*$ is countable. □□

Proposition 9.3 $\lambda(E_\infty \cap [\xi,\eta]) = 0$.

Proof Let $N \geqq 1$ and $\Lambda > 0$ be as in Proposition 9.2 and put $\omega = (d-c)\Lambda^{-1}$.

Lemma 9.11 If $n \geqq N$ and J is a regular component of $\tilde{E}_n$ then

$|\hat{J}| \geq \omega|J|$.

Proof By Proposition 9.2 we have $\frac{|f^n(\hat{J})|}{|f^n(J)|} \leq \Lambda \frac{|\hat{J}|}{|J|}$, and $|f^n(J)| \leq 1$, $|f^n(\hat{J})| = |[c,d]| = d-c$. Thus $|\hat{J}| \geq (d-c)\Lambda^{-1}|J| = \omega|J|$. ◻◻

Lemma 9.12 For each $n \geq N$ we have $\lambda(E^*_\infty \cap [\xi,\eta]) \leq (1-\omega)\lambda(\tilde{E}_n)$.

Proof Let $n \geq N$ and J be a component of $\tilde{E}_n$. Since $E^*_\infty \cap [\xi,\eta] \subset \tilde{E}_n$ it is enough to show that $\lambda(E^*_\infty \cap [\xi,\eta] \cap J) \leq (1-\omega)|J|$. This holds trivially if $E^*_\infty \cap [\xi,\eta] \cap J = \phi$ and so we can assume there exists $x \in E^*_\infty \cap [\xi,\eta] \cap J$. (Note that we then have $J = E_n(x)$.) Let

$$\ell = \min\{ k \geq n : E_k(x) \text{ is a regular component of } E_k \} ;$$

it is easy to see that $(J - E_\ell(x)) \cap E_\infty = \phi$ and also $\widehat{E_\ell(x)} \cap E_\infty = \phi$, and therefore $E^*_\infty \cap [\xi,\eta] \cap J \subset E_\ell(x) - \widehat{E_\ell(x)}$. But $E_\ell(x) \in \tilde{E}_\ell$ (because $x \in [\xi,\eta]$) and $\ell \geq n \geq N$; thus by Lemma 9.11 we have $|\widehat{E_\ell(x)}| \geq \omega|E_\ell(x)|$ and hence

$$\lambda(E^*_\infty \cap [\xi,\eta] \cap J) \leq |E_\ell(x)| - |\widehat{E_\ell(x)}| \leq (1-\omega)|E_\ell(x)| \leq (1-\omega)|J| . \quad ◻◻$$

Now $\{\tilde{E}_n\}_{n \geq 1}$ is a decreasing sequence with $\bigcap_{n \geq 1} \tilde{E}_n = E_\infty \cap [\xi,\eta]$; thus by Lemma 9.12 we have

$$\lambda(E^*_\infty \cap [\xi,\eta]) \leq (1-\omega) \lim_{n \to \infty} \lambda(\tilde{E}_n) = (1-\omega)\lambda(E_\infty \cap [\xi,\eta]) .$$

But Lemma 9.10 implies that $\lambda(E^*_\infty \cap [\xi,\eta]) = \lambda(E_\infty \cap [\xi,\eta])$, and hence $\lambda(E_\infty \cap [\xi,\eta]) = 0$. ◻◻

We now know that $\lambda(E_\infty \cap [\xi,\eta]) = 0$. However, given any $\varepsilon > 0$ we could have chosen ξ and η so that $[\varepsilon, 1-\varepsilon] \subset [\xi,\eta]$, and thus we must in fact have $\lambda(E_\infty) = 0$. This then completes the proof of Theorem 9.3(2). ◻◻

Proof of Theorem 9.2 The key result here is the following:

Proposition 9.4 Let $f \in S_R^{**} \cap S_3$ and let $\Gamma = ([c,d],f^m)$ be the simple reduction of smallest order satisfying (6.5). Then $\lambda(C_\Gamma^*) = 1$, where $C_\Gamma^* = \bigcap_{\varepsilon>0} C_\Gamma(\varepsilon)$ and for each $\varepsilon > 0$

$$C_\Gamma(\varepsilon) = \{ x \in [0,1] : f^k(x) \in (c-\varepsilon, d+\varepsilon) \text{ for some } k \geqq 0 \} .$$

Proof We leave this to the end. ◻◻

Now let $f \in S_R^{**}$ with $|P_s(f)| = 1$, and let $[x] \in P_s(f)$. From the proof of Theorem 5.2 we know that $V^n f \in S_1$ for some $n \geqq 1$; let $\ell = \min\{ n \geqq 1 : V^n f \in S_1 \}$. The proof of Theorem 9.1(1) will give us that $\lambda(A([x],f)) = 1$ (and hence $\lambda(A^*([x],f)) = 1$) unless $V^k f$ is a non-proper element of S_3 for some $k \geqq 0$; we can thus assume that this is the case. But then $V^{k+1} f \in S_1$ and hence $k = \ell-1$; moreover (as in the proof of Lemma 9.3(3)) we must have $(V^\ell f)(z) < z$ for all $z \in (0,\varphi(V^\ell f)]$ and so $\lim_{n\to\infty} (V^\ell f)^n(z) = 0$ for all $z \in [0,1]$. Using this and applying Proposition 9.4 to $V^{\ell-1} f$ we obtain that $V^{\ell-1} f \in G^*$, where

$$G^* = \{ g \in S : |P_s(g)| = 1 \text{ and if } [z] \in P_s(g) \text{ then } \lambda(A^*([z],g)) = 1 \} .$$

A slight modification of Lemma 9.2 now gives us that $V^n f \in G^*$ for all $0 \leqq n \leqq \ell-1$, and so in particular $f \in G^*$. ◻◻

Proof of Proposition 9.4 If f is proper then the result follows from Theorem 9.3(2). We can thus assume that f is not proper, and so by Lemma 9.1 $[e_\Gamma]$ is one-sided stable. Note that $\alpha < c < d < \beta$ (since $f \in S_3$).

Lemma 9.13 For each $\varepsilon > 0$ we can find a periodic point of f in the interval $(c-\varepsilon,c)$.

Proof Without loss of generality we can assume that $\varepsilon < c-\alpha$. We have $c \in [0,1] - (\mathrm{int}(D) \cup \mathrm{int}(E))$ (where D and E are defined in the proof of Propositions 6.2 and 6.3), and therefore as in Lemma 6.9 there exists $k \geqq 1$ such that $f^k((c-\varepsilon,c]) \supset [\alpha,\beta]$. But $[c-\varepsilon,c] \subset (\alpha,\beta)$ and so there must exist $z \in (c-\varepsilon,c)$ with $f^k(z) = z$. ▫▫

Now fix $\varepsilon > 0$; Lemma 9.13 and the fact that $f(c) = f(d)$ (cf. the proof of Lemma 9.6) imply that we can find $a \in (c-\varepsilon,c)$ and $b \in (d,d+\varepsilon)$ so that a is a periodic point of f , $a \notin [e_\Gamma]$ and $f(a) = f(b)$. For $n \geqq 1$ put

$$F_n = \{ x \in [0,1] : f^k(x) \notin (a,b) \text{ for } k = 0,1,\ldots,n-1 \} ,$$

and let $F_\infty = \bigcap_{n \geqq 1} F_n$. Let p be the period of a .

Lemma 9.14 (1) $\lim_{n\to\infty} \theta_n = 0$, where θ_n is the length of the largest interval contained in F_n .

(2) $f^p(a) = f^p(b) = a$.

(3) $|(f^p)'(a)| > 1$.

(4) There exists $\zeta > 0$ such that $|f^k([a,\varphi])| \geqq \zeta$ and $|f^k([\varphi,b])| \geqq \zeta$ for all $k \geqq 0$.

Proof (1): We have $F_n \subset E_n$ and thus $\theta_n \leqq \tau_n$. Hence $\lim_{n\to\infty} \theta_n = 0$ follows from Lemma 9.4.

(2): This is clear since $f(a) = f(b)$ and $f^p(a) = a$.

(3): a is unstable because $[a] \neq [e_\Gamma]$ and $P_s(f) = \{[e_\Gamma]\}$; thus (as noted in the proof of Lemma 9.1) we must have $|(f^p)'(a)| > 1$.

(4): We have $f^{pm}([a,\varphi]) \supset f^{pm}([a,c]) \supset [a,e_\Gamma]$, $f^{pm}([\varphi,b]) \supset f^{pm}([d,b]) \supset [a,e_\Gamma]$ and $f^{pm}([a,e_\Gamma]) \supset [a,e_\Gamma]$. Thus for all $\ell \geqq 0$ we have $f^{\ell pm}([a,\varphi]) \supset [a,e_\Gamma]$ and $f^{\ell pm}([\varphi,b]) \supset [a,e_\Gamma]$, and the result is an immediate consequence of this. ☐☐

Now using the four properties given in Lemma 9.14 we obtain $\lambda(F_\infty) = 0$ in exactly the same way as in the proof that $\lambda(E_\infty) = 0$. But $[0,1] - C_\Gamma(\varepsilon) \subset F_\infty$ and hence $\lambda(C_\Gamma(\varepsilon)) = 1$; since this holds for all $\varepsilon > 0$ we thus have $\lambda(C_\Gamma^*) = 1$. ☐☐

Notes: Theorems 9.1 and 9.2 are due to Guckenheimer and Misiurewicz (Guckenheimer (1979), Misiurewicz (1980)). The proofs given here are adapted from Guckenheimer (1979). Guckenheimer in fact asserts that $\lambda(A([x],f)) = 1$ when $[x]$ is one-sided stable; however, in the proof of this it is claimed that $A^*([x],f) = A([x],f)$, which is not the case.

10. OCCURRENCE OF THE DIFFERENT TYPES OF BEHAVIOUR

Let $S_R^{**} = \{ f \in S_R^* : f(z) > z$ for all $z \in (0,\varphi] \}$; Theorem 5.1 tells us that if $f \in S_R^{**}$ then exactly one of (5.6), (5.7) and (5.8) holds. In this section we will show that in any "reasonable" one-parameter family of functions from S_R^{**} (such as $f_\mu(x) = \mu x(1-x)$, $2 < \mu \leqq 4$, $f_\mu(x) = \sin(\mu x)$, $\frac{\pi}{2} < \mu \leqq \pi$ and $f_\mu(x) = \mu x \exp(1-\mu x)$, $\mu > 1$) there are infinitely many examples of each of the three types.

Once we have examples of functions satisfying (5.6), (5.7) and (5.8) then, in the more general setting of Theorem 5.2, we automatically have examples of functions for which (5.11), (5.13) and (5.14) holds. We will show that functions satisfying (5.12) are very easy to find.

We have already seen how examples of functions satisfying (5.6) and (5.7) can be constructed, viz:

Proposition 10.1 (1) Let $f \in S_R^{**}$ have a continuous derivative in $(0,1)$ and suppose that φ is periodic. Then f satisfies (5.6).

(2) Let $f \in S_R^{**}$ be such that $f^n(\varphi) = \beta$ for some $n \geqq 2$ (where β is the unique fixed point of f in $(\varphi,1)$). Then f satisfies (5.7).

Proof (1): Let φ have period m ; since φ is periodic we have $[\varphi] \subset (0,1)$ and thus f^m is continuously differentiable in a neighbourhood of φ . Therefore φ is stable and so by Theorem 5.1 we have that f satisfies (5.6).

(2): As noted after the proof of Proposition 4.3 β is unstable and so by Proposition 4.3 we have $|P_s(f)| = 0$. Thus either (5.7) or (5.8) holds. But $\varphi \in A(f)$ and hence Proposition 5.3 implies that (5.7) holds.

□□

The next result shows that in a "reasonable" one-parameter family from $\mathcal{S}$ there are infinitely many functions satisfying the hypotheses of Proposition 10.1. If $\{f_\mu\}_{\mu\in I}$ is a one-parameter family then we will write φ_μ , β_μ and α_μ instead of $\varphi(f_\mu)$, $\beta(f_\mu)$ and $\alpha(f_\mu)$; also we write f_μ^n instead of $(f_\mu)^n$.

Proposition 10.2 Let $I \subset \mathbb{R}$ be an interval and for each $\mu \in I$ let $f_\mu \in \mathcal{S}$ with $f_\mu(z) > z$ for all $z \in (0,\varphi_\mu]$; let $F : [0,1]\times I \to [0,1]$ be given by $F(x,\mu) = f_\mu(x)$. Suppose that:

(10.1) F is continuous.

(10.2) There exists $u \in I$ with $f_u^2(\varphi_u) \geqq \varphi_u$.

(10.3) For each $n \geqq 2$ there exists $v \in I$ with

$$f_v^2(\varphi_v) \leqq f_v^3(\varphi_v) \leqq \cdots \leqq f_v^n(\varphi_v) \leqq \varphi_v .$$

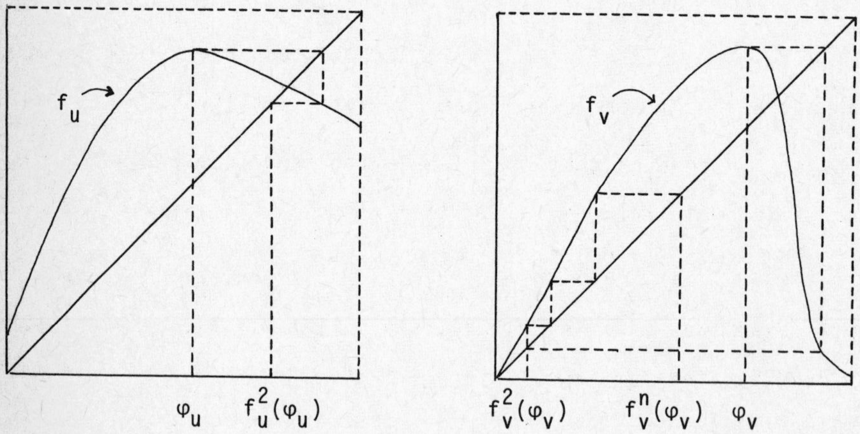

(10.2) (10.3)

Then $\{ \mu \in I : \varphi_\mu$ is periodic $\}$ and

$\{ \mu \in I : f_\mu^n(\varphi_\mu) = \beta_\mu$ for some $n \geqq 2 \}$ are both infinite.

Remark: If there exists $w \in I$ with $f_w(\varphi_w) = 1$, $f_w(1) = 0$ and $f_w(0) = 0$ then (10.3) is trivially satisfied (taking $v = w$ for all $n \geqq 2$). Thus $f_\mu(x) = \mu x(1-x)$, $2 < \mu \leqq 4$, and $f_\mu(x) = \sin(\mu x)$, $\frac{\pi}{2} < \mu \leqq \pi$, both satisfy (10.3). The family $f_\mu(x) = \mu x \exp(1-\mu x)$, $\mu > 1$, also satisfies (10.3) because here we have $\varphi_\mu = \frac{1}{\mu}$ and $f_\mu^{k+1}(\varphi_\mu) \leqq \mu^k \exp(k-\mu)$ for all $\mu > 1$, $k \geqq 1$. Moreover, it is easy to see that all three families satisfy (10.1) and (10.2).

Before proving Proposition 10.2 we give a lemma which will be needed in all the proofs in this section.

Lemma 10.1 Let $I \subset \mathbb{R}$ be an interval and for each $\mu \in I$ let $f_\mu \in S$; let $F : [0,1]\times I \longrightarrow [0,1]$ be given by $F(x,\mu) = f_\mu(x)$, and suppose that F is continuous. Then we have:

(1) φ_μ is a continuous function of μ .

(2) If $f_\mu(\varphi_\mu) > \varphi_\mu$ for all $\mu \in I$ then β_μ and α_μ are both continuous functions of μ .

Proof This is left as an elementary exercise for the reader. ◻◻

Proof of Proposition 10.2 $f_u^2(\varphi_u) \geqq \varphi_u$ implies that $f_u([\varphi_u,f_u(\varphi_u)]) \subset [\varphi_u,f_u(\varphi_u)]$ and thus $f_u^3(\varphi_u) > \beta_u > \varphi_u$. Now fix $n \geqq 3$ and let

$$L_n = \{ \mu \in I : f_\mu^2(\varphi_\mu) \leqq f_\mu^3(\varphi_\mu) \leqq \cdots \leqq f_\mu^n(\varphi_\mu) \leqq \varphi_\mu \} ;$$

by Lemma 10.1, (10.1) and (10.3) we have L_n is a closed, non-empty subset of I , and $u \notin L_n$. Let $\nu \in L_n$ be such that $|\nu-u| \leqq |\mu-u|$ for all $\mu \in L_n$; then either $f_\nu^n(\varphi_\nu) = \varphi_\nu$ or $f_\nu^k(\varphi_\nu) = f_\nu^{k+1}(\varphi_\nu)$ for some $2 \leqq k < n$. But the latter can only hold if $f_\nu(0) = 0$ and

$f_\nu^j(\varphi_\nu) = 0$ for all $j \geqq 2$, and this is not possible (because it would imply that $\nu \in \operatorname{int}(L_n)$). Thus $f_\nu^n(\varphi_\nu) = \varphi_\nu$, and since $\nu \in L_n$ it follows that φ_ν has period n . Therefore for each $n \geqq 3$ there exists $\nu \in I$ such that φ_ν is periodic with period n and so in particular $\{ \mu \in I : \varphi_\mu$ is periodic $\}$ is infinite.

Now exactly the same proof, but using

$$L_n^* = \{ \mu \in I : f_\mu^2(\varphi_\mu) \leqq f_\mu^3(\varphi_\mu) \leqq \cdots \leqq f_\mu^n(\varphi_\mu) \leqq \alpha_\mu \}$$

instead of L_n , shows that for each $n \geqq 3$ there exists $\nu \in I$ such that $f^n(\varphi_\nu) = \alpha_\nu$, and this implies that $\{ \mu \in I : f_\mu^n(\varphi_\mu) = \beta_\mu$ for some $n \geqq 2 \}$ is infinite. (Note that L_n^* is non-empty because $L_{n+1} \subset L_n^*$.) ☐☐

Note: For the functions f_ν constructed in Proposition 10.2 the iterates of φ_ν have the following simple structure:

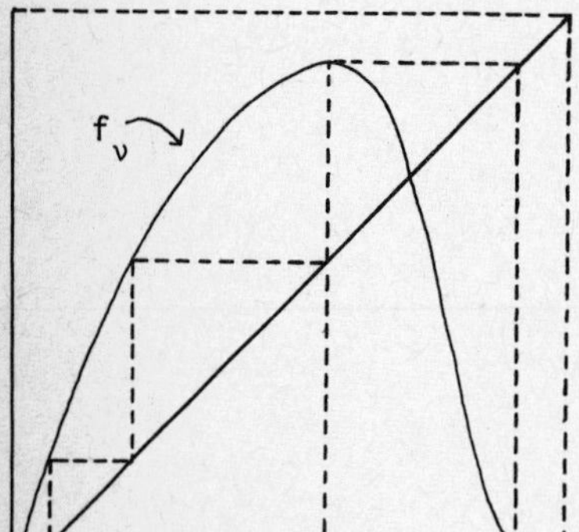

φ_ν periodic

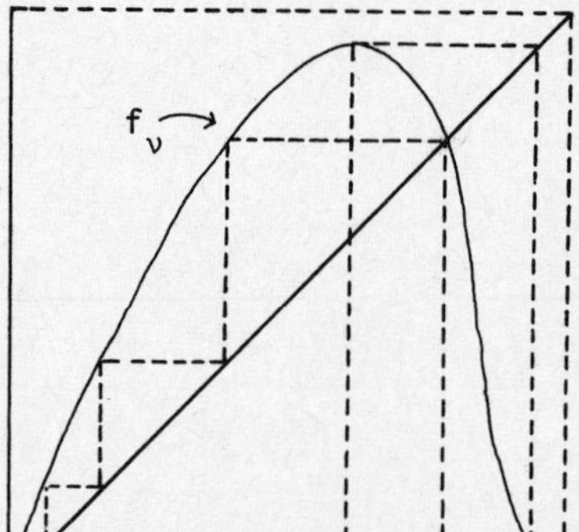

$f_\nu^n(\varphi_\nu) = \beta_\nu$ for some n

We now consider the problem of finding functions which satisfy (5.8). If $f \in S_R^{**}$ then the proof of Theorem 5.2 tells us that (5.8) holds if and only if $V^n f \in S_2 \cup S_3$ for all $n \geqq 0$; we start by looking for functions f for which $V^n f \in S_2$ for all $n \geqq 0$. Recall that $S_2 = \{ f \in S : f(\varphi) > \varphi$ and $f^2(\varphi) \geqq \alpha \}$ and that if $f \in S_2$ then $Vf = Tf$, where $Tf \in S$ is given by

$$(Tf)(x) = \frac{1}{\beta-\alpha} \left(\beta - f^2(\beta-(\beta-\alpha)x) \right) .$$

Theorem 10.1 Let $I \subset \mathbb{R}$ be an interval and for each $\mu \in I$ let $f_\mu \in S$ with $f_\mu(\varphi_\mu) > \varphi_\mu$; let $F : [0,1]\times I \rightarrow [0,1]$ be given by $F(x,\mu) = f_\mu(x)$. Suppose that F is continuous in $[0,1]\times I$ and continuously differentiable in $(0,1)\times \mathrm{int}(I)$; suppose also there exist $u, v \in \mathrm{int}(I)$ with $f_u^2(\varphi_u) \geqq \varphi_u$ and $f_v^2(\varphi_v) \leqq \alpha_v$. Then there exists $w \in I$ such that $V^n f_w \in S_2$ for all $n \geqq 0$.

Note: It is easy to check that the three families of functions mentioned at the beginning of this section all satisfy the hypotheses of Theorem 10.1.

Proof Without loss of generality we can assume that $u < v$. It is convenient to make the following definition: Let $a, b \in \mathbb{R}$ with $a < b$ and for each $\mu \in [a,b]$ let $g_\mu \in S$. We call $\{g_\mu\}_{\mu\in[a,b]}$ *suitable* if:

(10.4) $g_\mu(\varphi(g_\mu)) > \varphi(g_\mu)$ for each $\mu \in [a,b]$.

(10.5) $g_a^2(\varphi(g_a)) = \varphi(g_a)$.

(10.6) $g_b^2(\varphi(g_b)) = \alpha(g_b)$.

(10.7) $\alpha(g_\mu) < g_\mu^2(\varphi(g_\mu)) < \varphi(g_\mu)$ for all $\mu \in (a,b)$.

(10.8) $G : [0,1]\times[a,b] \longrightarrow [0,1]$ is continuous, where $G(x,\mu) = g_\mu(x)$.

(10.9) If for some $\nu \in [a,b)$, $m \geqq 2$ and $\varepsilon > 0$ we have $g_\nu^m(\varphi(g_\nu)) = \varphi(g_\nu)$ and $g_\mu^m(\varphi(g_\mu)) < \varphi(g_\mu)$ for all $\mu \in (\nu,\nu+\varepsilon)$ (resp. $g_\mu^m(\varphi(g_\mu)) > \varphi(g_\mu)$ for all $\mu \in (\nu,\nu+\varepsilon)$) then there exists $\delta > 0$ such that $g_\mu^{2m}(\varphi(g_\mu)) \leqq \varphi(g_\mu)$ for all $\mu \in (\nu,\nu+\delta)$ (resp. $g_\mu^{2m}(\varphi(g_\mu)) \geqq \varphi(g_\mu)$ for all $\mu \in (\nu,\nu+\delta)$).

Note that if $\{g_\mu\}_{\mu\in[a,b]}$ is suitable then $g_\mu \in \mathcal{S}_2$ for each $\mu \in [a,b]$.

Theorem 10.1 is a straightforward consequence of the two facts given in the next result.

Proposition 10.3 (1) If $\{g_\mu\}_{\mu\in[a,b]}$ is suitable then there exist $a < c < d < b$ such that $\{\mathcal{T}g_\mu\}_{\mu\in[c,d]}$ is suitable.

(2) If $\{f_\mu\}_{\mu\in I}$ satisfies the hypotheses of Theorem 10.1 (and $u < v$) then there exist $u \leqq a < b \leqq v$ so that $\{f_\mu\}_{\mu\in[a,b]}$ is suitable.

Proof Later. □□

Now let $\{g_\mu\}_{\mu\in[a,b]}$ be suitable; by continually applying Proposition 10.3 (1) we can find $a = a_0 < a_1 < \cdots < b_1 < b_0 = b$ such that for each $n \geqq 0$ we have $\mathcal{V}^n g_\mu \in \mathcal{S}_2$ for all $\mu \in [a_n,b_n]$ and $\{\mathcal{T}\mathcal{V}^n g_\mu\}_{\mu\in[a_{n+1},b_{n+1}]}$ is suitable. In particular, if $w \in \bigcap_{n\geqq 0} [a_n,b_n]$ then $\mathcal{V}^n g_w \in \mathcal{S}_2$ for all $n \geqq 0$. This, together with Proposition 10.3(2), clearly gives us Theorem 10.1. □□

Proof of Proposition 10.3(1) For $\mu \in [a,b]$ let $h_\mu = \mathcal{T}g_\mu$ and define $H : [0,1]\times[a,b] \longrightarrow [0,1]$ by $H(x,\mu) = h_\mu(x)$.

Lemma 10.2 H is continuous.

Proof We have

$$H(x,\mu) = (\beta(g_\mu)-\alpha(g_\mu))^{-1} \Big(\beta(g_\mu) - g_\mu^2(\beta(g_\mu)-(\beta(g_\mu)-\alpha(g_\mu))x) \Big) ,$$

and Lemma 10.1 gives us that $\beta(g_\mu)$ and $\alpha(g_\mu)$ are continuous functions of μ . Thus the continuity of H follows from (10.8). ▫▫

Lemma 10.3 $\{h_\mu\}_{\mu\in[a,b]}$ satisfies (10.9).

Proof Suppose for some $\nu \in [a,b)$, $m \geqq 2$ and $\varepsilon > 0$ we have $h_\nu^m(\varphi(h_\nu)) = \varphi(h_\nu)$ and $h_\mu^m(\varphi(h_\mu)) < \varphi(h_\mu)$ for all $\mu \in (\nu,\nu+\varepsilon)$. Then $g_\nu^{2m}(\varphi(g_\nu)) = \varphi(g_\nu)$ and $g_\mu^{2m}(\varphi(g_\mu)) > \varphi(g_\mu)$ for all $\mu \in (\nu,\nu+\varepsilon)$. There thus exists $\delta > 0$ so that $g_\mu^{4m}(\varphi(g_\mu)) \geqq \varphi(g_\mu)$ for all $\mu \in (\nu,\nu+\delta)$ and hence $h_\mu^{2m}(\varphi(h_\mu)) \leqq \varphi(h_\mu)$ for all $\mu \in (\nu,\nu+\delta)$. The other case follows in the same way. ▫▫

Note that we have $h_\mu(\varphi(h_\mu)) > \varphi(h_\mu)$ for all $\mu \in (a,b]$ (since $g_\mu^2(\varphi(g_\mu)) < \varphi(g_\mu)$ for all $\mu \in (a,b]$). Now $h_b^2(\varphi(h_b)) = 0$ (since $g_b^4(\varphi(g_b)) = g_b^2(\alpha(g_b)) = \beta(g_b)$) and there exists $\delta > 0$ such that $h_\mu^2(\varphi(h_\mu)) \geqq \varphi(h_\mu)$ for all $\mu \in (a,a+\delta)$. ($g_a^2(\varphi(g_a)) = \varphi(g_a)$ and $g_\mu^2(\varphi(g_\mu)) < \varphi(g_\mu)$ for all $\mu \in (a,b)$; thus by (10.9) there exists $\delta > 0$ such that $g_\mu^4(\varphi(g_\mu)) \leqq \varphi(g_\mu)$ for all $\mu \in (a,a+\delta)$, and hence $h_\mu^2(\varphi(h_\mu)) \geqq \varphi(h_\mu)$ for all $\mu \in (a,a+\delta)$.) But Lemma 10.1 implies that $\varphi(h_\mu)$, $\alpha(h_\mu)$ and $h_\mu^2(\varphi(h_\mu))$ are continuous functions of μ (in $(a,b]$), and thus if we define $c = \sup\{ \mu \in (a,b] : h_\mu^2(\varphi(h_\mu)) \geqq \varphi(h_\mu) \}$ and $d = \inf\{ \mu \in [c,b] : h_\mu^2(\varphi(h_\mu)) \leqq \alpha(h_\mu) \}$ then $a < c < d < b$, $h_c^2(\varphi(h_c)) = \varphi(h_c)$, $h_d^2(\varphi(h_d)) = \alpha(h_d)$ and $\alpha(h_\mu) < h_\mu^2(\varphi(h_\mu)) < \varphi(h_\mu)$

for all $\mu \in (c,d)$. This, together with Lemmas 10.2 and 10.3, gives us that $\{h_\mu\}_{\mu\in[c,d]}$ is suitable. ☐☐

Proof of Proposition 10.3(2) By Lemma 10.1 φ_μ , α_μ and $f^2_\mu(\varphi_\mu)$ are continuous functions of μ and thus if we put

$$a = \sup\{ \mu \in [u,v] : f^2_\mu(\varphi_\mu) \geqq \varphi_\mu \}$$

and

$$b = \inf\{ \mu \in [a,v] : f^2_\mu(\varphi_\mu) \leqq \alpha_\mu \}$$

then we have $u \leqq a < b \leqq v$, $f^2_a(\varphi_a) = \varphi_a$, $f^2_b(\varphi_b) = \alpha_b$ and $\alpha_\mu < f^2_\mu(\varphi_\mu) < \varphi_\mu$ for all $\mu \in (a,b)$. Hence $\{f_\mu\}_{\mu\in[a,b]}$ satisfies (10.4), (10.5), (10.6), (10.7) and (10.8). Now let $\nu \in [a,b)$, $m \geqq 2$ and $\varepsilon > 0$ and suppose that $f^m_\nu(\varphi_\nu) = \varphi_\nu$ and $f^m_\mu(\varphi_\mu) < \varphi_\mu$ for all $\mu \in (\nu,\nu+\varepsilon)$. Since $f^m_\nu(\varphi_\nu) = \varphi_\nu$ we have $(f^m_\nu)'(\varphi_\nu) = 0$ and thus (because F is continuously differentiable in $(0,1)\times\mathrm{int}(I)$) there exists $\eta > 0$ such that $|(f^m_\mu)'(z)| \leqq 1$ for all $\mu \in (\nu,\nu+\eta)$ and for all $z \in (\varphi_\nu-\eta,\varphi_\nu+\eta)$. $f^m_\mu(\varphi_\mu)$ and φ_μ are continuous functions of μ , and so we can find $\xi > 0$ such that $|\varphi_\mu-\varphi_\nu| < \eta$ and $|f^m_\mu(\varphi_\mu)-\varphi_\nu| < \eta$ for all $\mu \in (\nu,\nu+\xi)$. Put $\delta = \min\{\eta,\xi,\varepsilon\}$ and let $\mu \in (\nu,\nu+\delta)$; by the mean value theorem there exists $z \in (f^m_\mu(\varphi_\mu),\varphi_\mu)$ such that

$$|f^{2m}_\mu(\varphi_\mu)-f^m_\mu(\varphi_\mu)| = |(f^m_\mu)'(z)||f^m_\mu(\varphi_\mu)-\varphi_\mu| .$$

But then $z \in (\varphi_\nu-\eta,\varphi_\nu+\eta)$ and so $|(f^m_\mu)'(z)| \leqq 1$. Therefore

$$f^{2m}_\mu(\varphi_\mu)-f^m_\mu(\varphi_\mu) \leqq |f^m_\mu(\varphi_\mu)-\varphi_\mu| = \varphi_\mu-f^m_\mu(\varphi_\mu) ;$$

i.e. $f^{2m}_\mu(\varphi_\mu) \leqq \varphi_\mu$. If $f^m_\nu(\varphi_\nu) = \varphi_\nu$ and $f^m_\mu(\varphi_\mu) > \varphi_\mu$ for all $\mu \in (\nu,\nu+\varepsilon)$ then in the same way we can find $\delta > 0$ so that

$f^{2m}(\varphi_\mu) \geqq \varphi_\mu$ for all $\mu \in (\nu,\nu+\delta)$. Thus $\{f_\mu\}_{\mu\in[a,b]}$ is suitable. ▫▫

Let $\{f_\mu\}_{\mu\in I}$ satisfy the hypotheses of Theorem 10.1 (with $u < v$). The proof of this result gives us $\{a_n\}_{n\geqq 0}$ and $\{b_n\}_{n\geqq 0}$ with $u \leqq a_0 < a_1 < a_2 < \cdots < b_2 < b_1 < b_0 \leqq v$ such that for each $n \geqq 0$ we have $V^n f_\mu \in S_2$ for all $\mu \in [a_n,b_n]$ and $\{TV^n f_\mu\}_{\mu\in[a_{n+1},b_{n+1}]}$ is suitable. Then $V^n f_w \in S_2$ for all $n \geqq 0$ for each $w \in [a_\infty,b_\infty]$, where $a_\infty = \lim_{n\to\infty} a_n$, $b_\infty = \lim_{n\to\infty} b_n$. In any "well-behaved" family (such as the three families mentioned at the beginning of this section) we will have $a_\infty = b_\infty$, and this construction will give us the only point $w \in I$ for which $V^n f_w \in S_2$ for all $n \geqq 0$. Note that by (10.5) we have $\varphi_{a_n} \in \mathrm{Per}(2^{n+1}, f_{a_n})$ for each $n \geqq 1$.

The construction used in the proof of Theorem 10.1 comes from a theory initiated by Feigenbaum and which deals with the quantitative behaviour of one-parameter families. Feigenbaum discovered that the successive bifurcations in any reasonable family from S exhibit a remarkable quantitative universality. (For instance: In any family such as one of our three examples $\lim_{n\to\infty} \frac{a_n - a_{n-1}}{a_{n+1} - a_n} = 4.66920..$ is true.) The mathematics needed to explain this kind of phenomena is unfortunately way beyond what we have used in these notes. The interested reader is recommended to look at Feigenbaum (1978), (1979), Collet and Eckmann (1980), Collet, Eckmann and Lanford (1980) and Lanford (1980).

For the family $f_\mu(x) = \mu x(1-x)$, $2 < \mu \leq 4$, we have $3.5699456 < a_\infty = b_\infty < 3.5699457$, and it was as an approximation to a_∞

that the parameter value 3.569946 was chosen in the introduction.

At the beginning of the section we claimed that in a "reasonable" one-parameter family from S_R^{**} there are infinitely many functions for which (5.8) holds. Theorem 10.1 (usually) gives us only one such function and so we are still a long way from substantiating our claim. However, before continuing with this let us consider for a moment the more general situation of Theorem 5.2. If $f \in S$ with $f(\varphi) > \varphi$ and $f^2(\varphi) \geqq \gamma$ then Theorem 5.2 tells us that exactly one of (5.11), (5.12), (5.13) and (5.14) holds. We thus have the question of whether each of these four types actually occurs. Now (5.6), (5.7) and (5.8) are special cases of (respectively) (5.11), (5.13) and (5.14), and so the results already obtained in this section show that the types satisfying (5.11), (5.13) and (5.14) do occur. Functions satisfying (5.12) are also not difficult to find: Let $f \in S$ be as in the following picture:

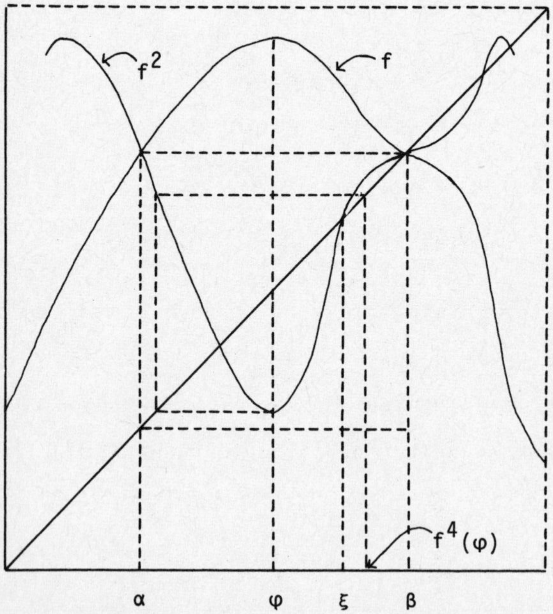

Thus $f \in S_2$, f^2 has a fixed point $\xi \in (\varphi,\beta)$ and $f^4(\varphi) \in (\xi,\beta)$. Then $Vf = Tf \in S_6$ and so Proposition 6.4 gives us that (5.12) holds. More generally we have that if $f \in S$ with $f(\varphi) > \varphi$ and $f^2(\varphi) \geqq \gamma$ then (5.12) holds provided there exists a trap J such that $f^k(\varphi) \in \mathrm{int}(J)$ for some $k \geqq 1$. (If this is the case then clearly f does not satisfy (5.11) or (5.14). But $\varphi \in (f^k)^{-1}(\mathrm{int}(J)) \subset S(f)$ and so by Proposition 3.3 (5.13) cannot hold.)

We now return to the problem of finding functions which satisfy (5.8).

Theorem 10.2 Let $I \subset \mathbb{R}$ be an interval and for each $\mu \in I$ let $f_\mu \in S_R$ with $f_\mu(z) > z$ for all $z \in (0,\varphi_\mu]$; let $F : [0,1]\times I \to [0,1]$ be given by $F(x,\mu) = f_\mu(x)$. Suppose that F is continuous in $[0,1]\times I$ and continuously differentiable in $(0,1)\times\mathrm{int}(I)$; suppose also there exist $u, v \in \mathrm{int}(I)$ with $f_u^2(\varphi_u) \geqq \varphi_u$ and $f_v^2(\varphi_v) \leqq \alpha_v$. Then $\{ \mu \in I : V^n f_\mu \in S_2 \cup S_3$ for all $n \geqq 0 \}$ is uncountable.

Remark: Except for the additional assumption of having property R the hypotheses of Theorem 10.2 are the same as those of Theorem 10.1. They are thus satisfied by the families $f_\mu(x) = \mu x(1-x)$, $2 < \mu \leqq 4$, $f_\mu(x) = \sin(\mu x)$, $\frac{\pi}{2} < \mu \leqq \pi$, and $f_\mu(x) = \mu x \exp(1-\mu x)$, $\mu > 1$, and hence for such a family of functions from S_R^{**} we have that $\{ \mu \in I : f_\mu$ satisfies (5.8) $\}$ is uncountable.

Proof As in the proof of Theorem 10.1 it is convenient to introduce a special kind of family: Let $a, b \in \mathbb{R}$ with $a < b$ and for each $\mu \in [a,b]$ let $g_\mu \in S$. We call $\{g_\mu\}_{\mu\in[a,b]}$ *full* if:

(10.10) $g_\mu \in S_R$ for each $\mu \in [a,b]$.

(10.11) $g_\mu(0) = g_\mu(1) = 0$ for each $\mu \in [a,b]$.

(10.12) If $\mu \in (a,b)$ then $g_\mu(z) > z$ for all $z \in (0,\varphi(g_\mu)]$.

(10.13) $G : [0,1]\times[a,b] \to [0,1]$ is continuous, where $G(x,\mu) = g_\mu(x)$.

(10.14) Either $g_a(\varphi(g_a)) = \varphi(g_a)$ and $g_b(\varphi(g_b)) = 1$ or $g_a(\varphi(g_a)) = 1$ and $g_b(\varphi(g_b)) = \varphi(g_b)$.

(10.15) If for some $\nu \in [a,b)$, $m \geq 1$ and $\varepsilon > 0$ we have $g_\nu^m(\varphi(g_\nu)) = \varphi(g_\nu)$ and $g_\mu^m(\varphi(g_\mu)) < \varphi(g_\mu)$ for all $\mu \in (\nu,\nu+\varepsilon)$ (resp. $g_\mu^m(\varphi(g_\mu)) > \varphi(g_\mu)$ for all $\mu \in (\nu,\nu+\varepsilon)$) then there exists $\delta > 0$ such that $g_\mu^{2m}(\varphi(g_\mu)) \leq \varphi(g_\mu)$ for all $\mu \in (\nu,\nu+\delta)$ (resp. $g_\mu^{2m}(\varphi(g_\mu)) \geq \varphi(g_\mu)$ for all $\mu \in (\nu,\nu+\delta)$).

(10.16) If for some $\nu \in (a,b]$, $m \geq 1$ and $\varepsilon > 0$ we have $g_\nu^m(\varphi(g_\nu)) = \varphi(g_\nu)$ and $g_\mu^m(\varphi(g_\mu)) < \varphi(g_\mu)$ for all $\mu \in (\nu-\varepsilon,\nu)$ (resp. $g_\mu^m(\varphi(g_\mu)) > \varphi(g_\mu)$ for all $\mu \in (\nu-\varepsilon,\nu)$) then there exists $\delta > 0$ such that $g_\mu^{2m}(\varphi(g_\mu)) \leq \varphi(g_\mu)$ for all $\mu \in (\nu-\delta,\nu)$ (resp. $g_\mu^{2m}(\varphi(g_\mu)) \geq \varphi(g_\mu)$ for all $\mu \in (\nu-\delta,\nu)$).

Theorem 10.2 is a straightforward consequence of the two facts given in the next result. For $m > 1$ let

$$S_3(m) = \{ f \in S_3 : \text{the simple reduction of smallest order satisfying (6.5) has order } m \} .$$

Proposition 10.4 (1) If $\{g_\mu\}_{\mu\in[a,b]}$ is full and $m \geq 3$ is prime then there exist $a < c < d < b$ such that $g_\mu \in S_3(m)$ for all $\mu \in [c,d]$ and $\{Vg_\mu\}_{\mu\in[c,d]}$ is full.

(2) If $\{f_\mu\}_{\mu\in I}$ satisfies the hypotheses of Theorem 10.2 then there

exist $a, b \in I$ with $a < b$ such that $f_\mu \in S_2$ for all $\mu \in [a,b]$ and $\{Tf_\mu\}_{\mu\in[a,b]}$ is full.

Proof Later. ⊞

Now let $\{f_\mu\}_{\mu\in I}$ satisfy the hypotheses of Theorem 10.2 and let $\{m_n\}_{n\geqq 1}$ be a sequence of primes with $m_n \geqq 3$ for each $n \geqq 1$. From Proposition 10.4(2) there exist $a, b \in I$ with $a < b$ such that $f_\mu \in S_2$ for all $\mu \in [a,b]$ and $\{Vf_\mu\}_{\mu\in[a,b]}$ is full. Thus by continually applying Proposition 10.4(1) we can find $a = a_0 < a_1 < \cdots < b_1 < b_0 = b$ such that for each $n \geqq 1$ we have $V^n f_\mu \in S_3(m_n)$ for all $\mu \in [a_n,b_n]$ and $\{V^{n+1}f_\mu\}_{\mu\in[a_{n+1},b_{n+1}]}$ is full. In particular, if $w \in \bigcap_{n\geqq 0} [a_n,b_n]$ then $f_w \in S_2$ and $V^n f_w \in S_3(m_n)$ for all $n \geqq 1$. This clearly gives us Theorem 10.2. ⊞

Proof of Proposition 10.4(2) Without loss of generality we can assume that $u < v$. By Proposition 10.3(2) there then exist $u \leqq a < b \leqq v$ such that $\{f_\mu\}_{\mu\in[a,b]}$ is suitable, and it thus follows that $\{Tf_\mu\}_{\mu\in[a,b]}$ satisfies (10.10), (10.11), (10.13) and (10.14). Now the proof of Proposition 10.3(2) shows that $\{f_\mu\}_{\mu\in[a,b]}$ satisfies (10.15) and (10.16), and (as in the proof of Proposition 10.3(1)) this implies that $\{Tf_\mu\}_{\mu\in[a,b]}$ also satisfies (10.15) and (10.16). But (10.7) and Proposition 4.6 give us that if $\mu \in (a,b)$ then $f_\mu^2(z) < z$ for all $z \in [\varphi_\mu,\beta_\mu)$, and hence $\{Tf_\mu\}_{\mu\in[a,b]}$ satisfies (10.12). Putting this all together we have that $\{Tf_\mu\}_{\mu\in[a,b]}$ is full. ⊞

Proof of Proposition 10.4(1) Let $\Gamma = ([c,d],f^m)$ be a simple reduction of $([0,1],f)$; we call Γ *proper* when $f^m(\varphi) \in \begin{cases} (\varphi,d) & \text{if } e_\Gamma = c\,, \\ (c,\varphi) & \text{if } e_\Gamma = d\,. \end{cases}$

Note that if $f \in S_R$ and Γ is proper then by Proposition 4.6 we have $f(z) > z$ for all $z \in (c,\varphi]$ (resp. $f(z) < z$ for all $z \in [\varphi,d)$) if $e_\Gamma = c$ (resp. if $e_\Gamma = d$).

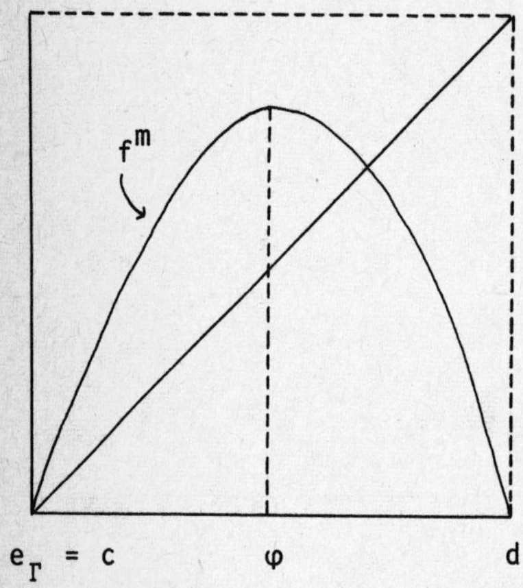

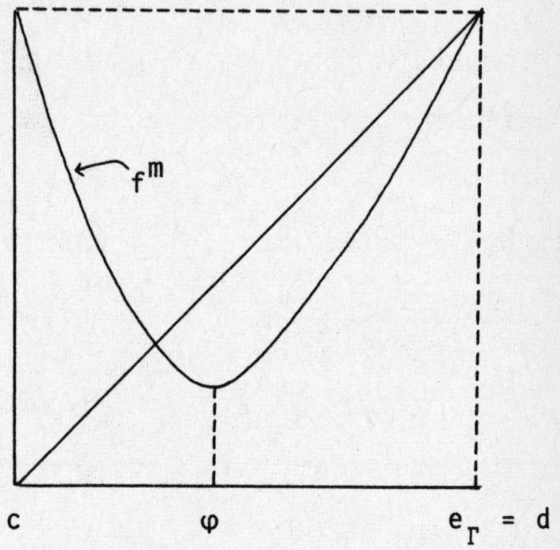

For $m > 1$ let

$$S_3^*(m) = \{ f \in S_3(m) : \text{the simple reduction of smallest order satisfying (6.5) is proper} \} .$$

Lemma 10.4 Let $\{g_\mu\}_{\mu\in[a,b]}$ be full and for $m > 1$ let $I_m = \{ \mu \in [a,b] : g_\mu \in S_3(m) \}$, $I_m^* = \{ \mu \in [a,b] : g_\mu \in S_3^*(m) \}$ and for $\mu \in I_m$ let $\Gamma_\mu = ([c_\mu,d_\mu],g_\mu^m)$ be the simple reduction of smallest order satisfying (6.5). Then:

(1) If $m \geqq 3$ is prime then I_m^* is an open subset of $[a,b]$.

(2) Let $m \geqq 3$ be prime, suppose that $I_m \neq \phi$ and let $[p,q] = \overline{J}$, where J is a (maximal connected) component of I_m^* . Then $g_p, g_q \in I_m$ and c_μ and d_μ are continuous functions of μ on $[p,q]$.

(3) Suppose $p \in [a,b)$ is such that $\varphi(g_p)$ is periodic with period m

(where $m \geqq 3$ is prime) and there exists $\varepsilon > 0$ such that $g_\mu^m(\varphi(g_\mu)) < \varphi(g_\mu)$ for all $\mu \in (p,p+\varepsilon)$. Then $I_m^* \neq \phi$ and p is the left-hand end-point of the closure of some component of I_m^* .

Proof We leave this to the end. ⊞

Now let $\{g_\mu\}_{\mu\in[a,b]}$ be full, suppose that $I_m^* \neq \phi$ and let $[p,q] = \overline{J}$, where J is a component of I_m^* . For $\mu \in [p,q]$ we have

$$(Vg_\mu)(x) = \begin{cases} \dfrac{1}{d_\mu - c_\mu}\Big(g_\mu^m((d_\mu - c_\mu)x + c_\mu) - c_\mu \Big) & \text{if } e_{\Gamma_\mu} = c_\mu , \\ \dfrac{1}{d_\mu - c_\mu}\Big(d_\mu - g_\mu^m(d_\mu - (d_\mu - c_\mu)x) \Big) & \text{if } e_{\Gamma_\mu} = d_\mu , \end{cases}$$

and it is clear that either $e_{\Gamma_\mu} = c_\mu$ for all $\mu \in [p,q]$ or $e_{\Gamma_\mu} = d_\mu$ for all $\mu \in [p,q]$. Thus if $H : [0,1]\times[p,q] \longrightarrow [0,1]$ is given by $H(x,\mu) = (Vg_\mu)(x)$ then Lemma 10.4(2) implies that H is continuous. Therefore $\{Vg_\mu\}_{\mu\in[p,q]}$ satisfies (10.10), (10.11), (10.12) and (10.13). Moreover, exactly as in the proof of Lemma 10.3 we also have that $\{Vg_\mu\}_{\mu\in[p,q]}$ satisfies (10.15) and (10.16). Hence $\{Vg_\mu\}_{\mu\in[a,b]}$ would be full if (10.14) held, and we will show that if $m \geqq 3$ is prime then it is possible to find a component J of I_m^* so that this is the case. Without loss of generality we can assume that $g_a(\varphi(g_a)) = \varphi(g_a)$ and $g_b(\varphi(g_b)) = 1$. Note that (because of (10.14) and (10.15)) $\{g_\mu\}_{\mu\in[a,b]}$ satisfies the hypotheses of Proposition 10.2. Thus if for $m \geqq 3$ we put $K_m = \{ \mu \in [a,b] : \varphi(g_\mu)$ is periodic with period $m \}$ then K_m is a closed, non-empty subset of $[a,b]$. Now fix $m \geqq 3$ prime and let $p = \max\{ \mu : \mu \in K_m \}$; we have $p < b$, and since $g_b^m(\varphi(g_b)) = g_b^{m-1}(1) = 0 < \varphi(g_b)$ it follows that $g_\mu^m(\varphi(g_\mu)) < \varphi(g_\mu)$ for all $\mu \in (p,b]$. (*Note:* If $g_\mu^m(\varphi(g_\mu)) = \varphi(g_\mu)$ then $\mu \in K_m$, because m is prime.) Thus by Lemma 10.4(3) we have $I_m^* \neq \phi$ and (since $g_b \notin S_3$)

there exists $q \in (p,b)$ such that (p,q) is a component of I_m^*. But $g_q \in I_m - I_m^*$ and so by Lemma 10.4(2) either $g_q^m(\varphi(g_q)) = \varphi(g_q)$ or $g_q^m(\varphi(g_q)) = c_{\Gamma_q}$. The former is not possible (because of our choice of p) and the latter implies that $(Vg_q)(\varphi(Vg_q)) = 1$. Hence $\{Vg_\mu\}_{\mu \in [p,q]}$ satisfies (10.14). ▫▫

Proof of Lemma 10.4 (1): Let $\nu \in I_m^*$ and without loss of generality let us assume that $e_{\Gamma_\nu} = c_\nu$. Since (c_ν,d_ν), $g_\nu((c_\nu,d_\nu)),\dots,g_\nu^{m-1}((c_\nu,d_\nu))$ are disjoint and $\varphi(g_\mu)$ is a continuous function of μ there exists $\xi > 0$ such that if $|\mu-\nu| < \xi$ then $\varphi(g_\mu) \notin g_\mu^k([c_\nu,d_\nu])$ for all $k = 1,\dots,m-1$; this implies that $\varphi(g_\mu)$ is the only turning point of g_μ^m in $[c_\nu,d_\nu]$. Now $c_\nu < g_\nu^{2m}(\varphi(g_\nu)) < \varphi(g_\nu)$ (because Γ_ν is proper) and $g_\nu^m(z) > z$ for all $z \in (c_\nu,\varphi(g_\nu)]$. Thus if we choose $w \in (c_\nu, g_\nu^{2m}(\varphi(g_\nu)))$ then we can find $0 < \eta \leqq \xi$ such that if $|\mu-\nu| < \eta$ then $\varphi(g_\mu) > w$, $g_\mu^m(z) > z$ for all $z \in (w,\varphi(g_\mu)]$, $g_\mu^m(\varphi(g_\mu)) < d_\nu$ and $g_\mu^{2m}(\varphi(g_\mu)) > w$. Let $|\mu-\nu| < \eta$; then (because g_μ^m is monotone on $(\varphi(g_\mu),d_\nu) \supset (\varphi(g_\mu),g_\mu^m(\varphi(g_\mu)))$) we have

$$g_\mu^m([w,g_\mu^m(\varphi(g_\mu))]) = [\min\{g_\mu^m(w),g_\mu^{2m}(\varphi(g_\mu))\},g_\mu^m(\varphi(g_\mu))]$$
$$\subset [w,g_\mu^m(\varphi(g_\mu))] .$$

Since $g_\nu \in S_3$ we have $g_\nu^2(\varphi(g_\nu)) < \alpha(g_\nu)$ and $\alpha(g_\nu) < c_\nu < d_\nu < \beta(g_\nu)$; we can thus find $0 < \delta \leqq \eta$ such that if $|\mu-\nu| < \delta$ then $g_\mu^2(\varphi(g_\mu)) < \alpha(g_\mu)$ and $\alpha(g_\mu) < c_\nu < d_\nu < \beta(g_\mu)$. Now let $|\mu-\nu| < \delta$; then $g_\mu^2(\varphi(g_\mu)) < \alpha(g_\mu)$, $\varphi(g_\mu) \in (w,g_\mu^m(\varphi(g_\mu)))$ and

$$g_\mu^m([w,g_\mu^m(\varphi(g_\mu))]) \subset [w,g_\mu^m(\varphi(g_\mu))] \subset [c_\nu,d_\nu] \subset (\alpha(g_\mu),\beta(g_\mu)) ;$$

it therefore follows that $\varphi(g_\mu)$ lies in a periodic component of $\mathrm{int}(\{ x \in [0,1] : g_\mu^k(x) \in [0,\alpha(g_\mu))$ for infinitely many $k \geqq 0 \})$.

Proposition 6.6 thus implies that $g_\mu \in S_3(m_\mu)$, where m_μ divides m . But m is prime and so $m_\mu = m$, i.e. $g_\mu \in S_3(m)$ whenever $|\mu-\nu| < \delta$. Finally, it is easy to see that c_μ and d_μ are continuous functions of μ in a neighbourhood of ν , and hence there exists $0 < \varepsilon \leqq \delta$ such that if $|\mu-\nu| < \varepsilon$ then $g_\mu \in S_3^*(m)$; i.e. I_m^* is open.

(2) and (3): These are similar to (1) and are left as an exercise for any reader who has managed to get this far. ◻◻

Notes: This section just gives some very elementary results about the behaviour of one-parameter families of functions from S . As we have already mentioned, the interested reader is recommended to look at Feigenbaum (1978), (1979), Collet and Eckmann (1980), Collet, Eckmann and Lanford (1980) and Lanford (1980) for further information concerning this subject.

REFERENCES

Allwright, D.J. (1978): *Hypergraphic functions and bifurcations in recurrence relations.* SIAM J. Appl. Math., 34, 687-691 .

Asmussen, M.A. and M.W. Feldman (1977): *Density dependent selection.* J. Theoretical Biology, 64, 603-618 .

Bowen, R. and J. Franks (1976): *The periodic points of maps of the disc and the interval.* Topology, 15, 337-343 .

Collet, P. and J.-P. Eckmann (1980): *Iterated maps on the interval as dynamical systems.* Progress in physics, Vol. 1 , Birkhäuser, Boston .

Collet, P., J.-P. Eckmann and O.E. Lanford (1980): *Universal properties of maps on an interval.* Comm. Math. Physics, 76, 211-254 .

Denjoy, A. (1932): *Sur les courbes définies par les équations différentielles à la surface du tore.* J. de Mathematiques Pures et Appliquées, 9. Série, 11, 333-375 .

Feigenbaum, M. (1978) and (1979): *Quantitative universality for a class of nonlinear transformations.* J. Stat. Physics, 19, 25-52; 21, 669-706 .

Guckenheimer, J. (1977): *On bifurcations of maps of the interval.* Inventiones Math., 39, 165-178 .

Guckenheimer, J. (1979): *Sensitive dependence to initial conditions for one-dimensional maps.* Comm. Math. Physics, 70, 133-160 .

Hille, E. (1976): *Ordinary differential equations in the complex domain.* Wiley-Interscience, New York .

Jonker, L. and D. Rand (1981): *Bifurcations in one dimension: 1. The nonwandering set.* Inventiones mathematicae, 62, 347-365; *2. A versal model for bifurcations.* Inventiones mathematicae, 63, 1-15 .

Lanford, O.E. (1980): *Smooth transformations of intervals.* Séminaire Bourbaki, 1980/81, n° 563, in: Springer Lecture Notes in Math., Vol. 901.

Li, T.Y. and J.A. Yorke (1975): *Period three implies chaos.* American Math. Monthly, 82, 985-992 .

May, R.B. (1975): *Biological populations obeying difference equations.* J. Theoretical Biology, 51, 511-524 .

May, R.B. (1976): *Simple mathematical models with very complicated dynamics.* Nature, 261, 459-467 .

May, R.B. and G. Oster (1976): *Bifurcations and dynamical complexity in simple ecological models.* American Naturalist, 110, 573-599 .

Metropolis, N., M.L. Stein and P.R. Stein (1973): *On finite limit sets for transformations on the unit interval.* J. Comb. Theory (A), 15, 25-44 .

Milnor, J. and W. Thurston (1976): *On iterated maps of the interval.* Preprint, Princeton .

Misiurewicz, M. (1980): *Absolutely continuous measures for certain maps of an interval.* Publ. Math. I.H.E.S.

Misiurewicz, M. and W. Szlenk (1977): *Entropy of piecewise monotone mappings.* Astérisque, 50, 299-310 .

Oster, G., A. Ipaktchi and S. Rocklin (1976): *Phenotypic structure and bifurcation behaviour of population models.* Theoret. Pop. Biol., 10, 365-382.

Parry, W. (1964): *Symbolic dynamics and transformations of the unit interval.* Transactions Am. Math. Soc., 122, 368-378 .

Rudin, W. (1964): *Principles of Mathematical Analysis.* McGraw-Hill, New York .

Šarkovskii, A.N. (1964): *Coexistence of cycles of a continuous map of a line into itself.* Ukranian Math. J., 16(1), 61-71 .

Schwarz, H.A. (1868): *Über einige Abbildungsaufgaben.* J. für reine und angewandte Mathematik, 70, 105-120 .

Singer, D. (1978): *Stable orbits and bifurcations of maps of the interval.* SIAM J. Appl. Math., 35, 260-267 .

Štefan, P. (1988): *A theorem of Šarkovskii on the existence of periodic orbits of continuous endomorphisms of the real line.* Comm. Math. Phys., 54, 237-248 .

Stein, P.R. and S. Ulam (1964): *Nonlinear transformation studies on electronic computers.* Rozprawy Matematyczne, 39, 401-484 .

Ulam, S. and J. von Neumann (1947): *On combination of stochastic and deterministic processes.* Bull. Am. Math. Soc., 53, 1120 .

INDEX

INDEX OF SYMBOLS